D1349477

Wireless Communications Security

For a complete listing of the *Artech House Universal Personal Communications Series,* turn to the back of this book. ·

Wireless Communications Security

Hideki Imai
Mohammad Ghulam Rahman
Kazukuni Kobara

ARTECH
HOUSE

BOSTON | LONDON
artechhouse.com

Library of Congress Cataloging-in-Publication Data
Imai, Hideki, 1943–
    Wireless communications security/Hideki Imai, Mohammad Ghulam Rahman,
Kazukuni Kobara.
        p. cm. — (universal personal communications)
    Includes bibliographical references and index.
    ISBN 1-58053-520-8 (alk. paper)
    1. Wireless communication systems—Security measures. 2. Mobile communica-
tion systems—Security measures. I. Rahman, Mohammad Ghulam. II. Kobara,
Kazukuni. III. Title IV. Artech House universal personal communications series.

    TK5103.2.I43 2005
    621.384—dc22   2005053075

British Library Cataloguing in Publication Data
Wireless communications security. — (Artech House universal personal
communications series)
    1. Wireless communication system—Security measures
    I. Imai, Hideki, 1943– II. Rahman, Mohammad Ghulam III. Kobara, Kazukuni
    621.3'82

    ISBN-10: 1-58053-520-8

Cover design by Igor Valdman

International Standard Book Number: 1-58053-520-8

10 9 8 7 6 5 4 3 2 1

Contents

Preface

Addressing the fast-growing need to integrate effective security features into wireless communication systems, this cutting-edge book offers you a broad overview of wireless security, allowing you to choose the methods and techniques most appropriate for your projects. You should gain a solid understanding of critical cryptography techniques, such as private/public key encryption, digital signatures, and block and stream ciphers. You will discover how to evaluate the impact of cryptography deployment on current wireless network architectures, and learn how to implement an effective cryptography architecture appropriate for your organization.

The book strives to offer the technical know-how needed to understand and work with the security concepts and techniques used for 2nd, 3rd, and 4th generation mobile networks. Moreover, this highly practical reference shows you how to implement an authentication technique for helping the user roam in different networks while maintaining the role symmetry of the message structure in intra domain and inter domain environments. Finally, the book discusses the Wireless Application Protocol (WAP), explaining how the protocol works and how to select appropriate cryptographic modules for this technology. We hope this book will help you comprehend the cryptographic solutions and how they should be applied properly to the wireless environment.

1

Introduction

Wireless communication systems are, by now, very popular all over the world. In some countries, like Japan and South Korea, the number of subscribers of cellular systems has passed that of public switched telephone. While wireless communication systems are very convenient for the user, their widespread use creates new challenges from a security point of view. It is physically far easier for an eavesdropper to intercept and illegitimately manipulate information if it is transmitted over the air rather than inside an optical fiber, for instance. Therefore, securing wireless systems has been a very active field both inside industry and academia.

In this book, we introduce recent developments in the field of wireless security. One of the distinguishing features of our work is that it is written from a cryptographic point of view (i.e., we mainly focus on the cryptographic solutions being available in the wireless environment and standards). Throughout this book, we explain what functionalities are provided with each cryptographic module and/or function, why and how they should be used, what level of resistance can be achieved with what size of parameters, and so on. As it is not rare that cryptographic modules and functions are used in improper ways, we believe our book is a valuable resource for technicians, engineers, and professionals working in the field of wireless communications. We require no previous knowledge of cryptography and all the

1

concepts used through the book are explained in a nontechnical but precise way.

Another highlight of this book, not present in other publications aimed at a similar audience, is its explanation of I-mode, the highly popular Japanese protocol for Web browsing with second generation cellular phones, and its security features.

We briefly introduce the main contents of each chapter.

In Chapter 2, we give a brief introduction to cryptography and its main techniques. The chapter has three main parts. In the first part, basic concepts of cryptography, such as types of cryptosystems, adversarial models, and security definitions are explained. In the second part, we present an introduction to symmetric key based cryptography with emphasis on block and stream ciphers. Finally, we conclude the chapter by explaining public key based cryptosystems (mostly encryption and signature protocols).

In Chapter 3, we give an introduction to wireless communication systems in general. We describe their characteristics, limiting and challenging aspects. We give a survey of which attacks are practical, to an eavesdropper, in this scenario and how they are reflected in security requirements for wireless networks.

In Chapter 4, we focus on wireless networking protocols different than cellular systems. We provide an overview of two of the mostly used standards: IEEE 802.11 (for wireless LAN) and Bluetooth (for wireless ad hoc networks).

In Chapter 5, we introduce the most popular cellular system of the second generation, the global system for mobile (GSM) communications. Also, we give a detailed explanation of the security of the Japanese I-mode (a service which enables Web browsing in standard second generation cell phones) and conclude the chapter by presenting an introduction to cellular digital packet data (CDPD), a service used in first generation system for the transmission of digital data. We point out strong and weak points of the security of those systems.

In Chapter 6, we study the security used in cellular systems of the third generation. We present the security protocols used in current systems, their analyses as well as a brief introduction to fourth generation cell systems and their requirements.

In Chapter 7, we explain the Wireless Application Protocol (WAP), which is an open specification that enables mobile users to have access to the Internet. WAP specifies both communication protocols and application environment so that it can work regardless of the underlying wireless networks and operating systems. We especially focus on the public-key infrastructure (PKI) model for WAP, wireless TLS (WTLS), WAP profiled TLS, WAP identity module (WIM), and explain their proper usage and the ideas behind them.

Cryptography may be an element of information security techniques, but cryptographic approach provides a strategic view of information security in general, because cryptology is the science treating secure information flow on the basis of strict definitions of security. We hope this book will help you obtain fundamental concepts and tools to solve any problem of wireless communications security.

2

Cryptography

2.1 Introduction

The primary goal of cryptography is to provide two people, usually called Alice and Bob, with a way to exchange secret messages, messages where no adversary, usually called Eve, can obtain any significant knowledge about their true meaning.

In this chapter, we review some important concepts of cryptography, which will be needed in the remaining chapters.

This chapter is divided in three main parts. First, we introduce basic concepts of cryptography, such as the definitions of symmetric/asymmetric encryption, and adversarial models. Then, in the second part, we introduce the basic ideas and techniques behind symmetric encryption schemes. In the third part, we concentrate on asymmetric encryption and digital signatures.

2.2 Basic Concepts

We now introduce the basic concepts and terminology of cryptology.

2.2.1 Ciphertext and Plaintext

To achieve secrecy over a public channel, Alice scrambles her message, usually known as plaintext, according to some preestablished rules, which are determined by a key shared in advance with Bob. After receiving the enciphered message, Bob unscrambles it, according to preagreed rules represented by his key, and recovers the plaintext. The processes of scrambling and unscrambling the message are called encryption and decryption, respectively. The scrambled text is called the ciphertext. This process is shown in Figure 2.1.

The ciphertext should provide Eve with no knowledge on the real message exchanged between Alice and Bob. Mathematically, we denote the encryption of a message m with a key ka by an encryption method $E(\ )$ by:

$$E_{ka}(m)$$

and the decryption of the same message m with a key kb by a decrypting algorithm $D(\ )$ by:

$$m = D_{kb}\left(E_{ka}(m)\right)$$

Encryption algorithms are also known as enciphering algorithms, and ciphers.

2.2.2   Types of Cryptosystems

A scheme, such as the one shown in Figure 2.1, is called a cryptosystem. If the keys used to encrypt and decrypt are the same, the cryptosystem is called a symmetric key one. If they are different, it is called an asymmetric key cryptosystem. The most common symmetric key cryptosystems are block ciphers and stream ciphers.

Block ciphers operate on the plaintext in groups of bits (blocks of bits) while stream ciphers operate on the plaintext bitwise. These ciphers are studied in more detail later in this chapter.

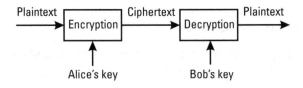

Figure 2.1 A general cryptosystem.

The most famous example of asymmetric key systems are public key cryptosystems (PKC). Whitfield Diffie and Martin Hellman introduced public-key cryptography in 1976 in a historical paper.

PKC uses two different keys in order to establish a secure communication, one public and other private. Anyone can encrypt a message using the public key, but only the owner of the private key can decrypt it. Through this scheme, two people that have never met before can establish a secure communication over an insecure channel. A practical construction for a PKC system was proposed by a group of researchers from MIT, the famous RSA.

At a first glance, asymmetrical cryptosystems seem to be more attractive than symmetrical ones, as they eliminate the problem of sharing a key in advance. However, from a computational point of view, asymmetrical cryptosystems are more expensive. To encrypt a long document with most of the available public key cryptosystems would be impractical. Thus, for example, asymmetrical cryptosystems are seldom employed in the case of mobile devices, which usually, have severe restrictions in their computational power.

The mostly employed solution is to combine symmetrical and asymmetrical cryptosystems as to make best use of their advantages. An asymmetric cryptosystem is used to share a small key, which in turn, will be used for encrypting longer messages with an asymmetric system.

2.2.3 Goals of a Cryptosystem

Besides providing secure exchanges of messages between two parties, modern cryptosystems are usually required to provide integrity, authenticity and non-repudiation. Here we briefly review these concepts:

- Secrecy means that an adversary is unable to learn a nonnegligible amount of information on the transmitted message;

- Integrity means that the recipient of the message, Bob, can be sure that it was not changed after being sent by Alice;

- Authenticity means that Bob can be sure that the received message was actually sent by Alice and no one else;

- Non-repudiation means that a sender is unable to send a message and later on claim that she has not sent it.

Sometimes, cryptosystems are required also to provide anonymity, that is, the identity of the parties using the system should be concealed from unauthorized people.

2.2.4 Security

In order to study the security of a cryptosystem, it is necessary first to specify what is available to an adversary who tries to break it. The following kinds of attacks are those most considered in the literature:

- *Ciphertext-only attack:* The adversary has access to a certain amount of encrypted data.

- *Known-plaintext attack:* The adversary has access not only to the ciphertext of a certain amount of data, but also to the data itself.

- *Chosen-plaintext attack:* The adversary can chose the data and its ciphertext which he is given access.

- *Adaptive chosen-plaintext attack:* The adversary can chose pairs of decrypted/encrypted data in an adaptive way, that is, she can modify her decisions about what pairs of plaintext/ciphertext she will receive based on previous received data.

- *Chosen-ciphertext attack:* The attacker has access to a machine which decrypts selected plaintexts. The goal of the attacker is to find the key used by the legal parties.

It is also important to remark that in cryptology, it is usually assumed that an adversary has access to a full description of the encryption and decryption algorithms. This is known as the Kerchoff Principle, that is, the security of a cryptosystem must rely solely on the secrecy of the key. It is wise to follow the Kirchoff Principle, since an enemy can always steal a cryptographic device and reverse engineer it or sometimes, reconstruct how the enciphering machine works solely from pairs of ciphertext/plaintext.

Since even though adversaries able to perform chosen-ciphertext attacks might look a bit too artificial at a first, one should bear in mind that an adversary can always steel/buy bits of plaintext and that many parts of official documents are standard (companies official names, for instance) and can be easily guessed.

2.3 Symmetric Encryption Schemes

In the following sections we provide basic explanations of the techniques and concepts related to symmetric encryption schemes, as defined earlier. After reviewing the one-time pad and explaining why it is impractical, we address some techniques that are important when designing block ciphers: substitution and permutation and confusion/diffusion. Then we give some examples of block ciphers and a brief explanation of stream ciphers, the most important classes of symmetrical encryption algorithms.

2.4 Perfect Secrecy: The One-Time Pad

A major problem when designing cryptosystems is how to define and measure the exact security of the system. A scientific study of the security of cryptosystems started with the work of Claude Shannon [1]. Shannon is also famous for being the first person to propose the use of Boolean algebra to analyze digital circuits, creating what is now called logical switching. In Shannon's later work, "A Mathematical Theory of Communication" [2], outlining what we now know as information theory, he described the measurement of information by binary digits representing yes-no alternatives, the fundamental basis of today's telecommunications.

What was most impressive is that, believe or not, Shannon did prove that a cipher proposed by Gilbert Vernan, a researcher at AT&T, is a perfect encryption scheme, secure even against all powerful adversaries. This scheme is known as the one-time pad. For simplicity, we will describe the one-time pad for binary messages {0,1}. Suppose that both the sender and the receiver possess a copy of a random sequence of 0's and 1's (the key). The sender, Alice, can encode a message by combining the message and the key using the exclusive-OR operation bitwise. An exclusive-OR operation, or XOR, will be denoted by the symbol + and is characterized by the following relations: $0 + 0 = 0$; $0 + 1 = 1$; $1 + 0 = 1$; $1 + 1 = 0$. In general words, when the two terms of the XOR operation are equal, the result will be zero. Also note that $a + b + b = a$.

Returning to the one-time-pad, Alice encodes the message M, by using a random key K, generating the ciphertext C in the following manner:

$$C = M + K$$

with all the operation taken bitwise.

Bob, the receiver, can easily recover the message by performing the following operation:

$$M = C + K$$

This process is illustrated in Figure 2.2.

We realize that the key and the plaintext must have the same length. The one-time pad is secure because the encrypted message, formed by the XOR of the message with the random key, is itself completely random. However, the key must be used only once. In order to communicate another message, say M'. Alice and Bob must share another key.

This restriction and the fact that the key must have the same length of the message make the one-time pad an almost impractical system. Its utility is restricted to communications that deserve the

$$
\begin{aligned}
M &= 0011100101 \\
K &= 1011010101 \\
\hline
C &= 1000110000
\end{aligned}
$$

The key must be used only once!
The key must have the same length of the message!

Figure 2.2 The one-time pad.

highest level of secrecy. For example, it was used in the communications between the White House, in Washington, D.C., and the Kremlin, in Moscow, on the famous red telephone.

Shannon not only proved that the one-time pad is perfectly secure, he also proved that any cryptosystem, in order to be perfectly secure, must have a key as large as the message to be transmitted. This makes unconditionally secure cryptosystems extremely expensive and difficult to use.

However, in a realistic scenario, it is common to assume that eavesdroppers are computationally restricted. In other words, it is common to assume that there are certain computational tasks that are difficult to be performed by an adversary. This idea is used in the so-called computationally secure cryptosystems. These are cryptosystems which have their security based on the difficulty of performing certain computations.

2.5 Block Ciphers

Block ciphers are the most used symmetric key enciphering schemes. Their name comes from the fact that block ciphers encrypt blocks of plaintext into blocks of ciphertext. Two important characteristics of a block cipher are the size of the key and the size of the blocks on which the cipher operates.

Before addressing the basic structure of block ciphers and explaining a few practical examples, we offer a review of important concepts and techniques that are used when designing block ciphers.

Two important techniques often used when designing symmetric key cryptosystems are permutation and substitution.

2.5.1 Permutation

In permutation, the letters of a message are simply rearranged, effectively generating an anagram. In this case of cryptosystem, the key will be the transposition rules. It is clear that for very short messages this system is insecure, because there are only a limited number of ways to rearrange a handful of letters. However, as the number of letters increase, so do the number of possibilities exponentially increase, making it impossible to recover the message without the scrambling rules. It is true that a random transposition offers a high security level, but the catch is that in order to be effective, the transposition rules must follow a straightforward system. It restricts many sets of possible keys (transpositions).

2.5.2 Substitution

Besides transposition, substitution is also another widely used technique to scramble a message into a ciphertext. In a substitution, each character of the message is substituted by another one from the same alphabet. Characters may be substituted individually or in groups. One trivial example is to pair letters of the alphabet at random, and then substitute each letter in the original letter by its partner. For example, the letters in the first line would be substituted by the letters in the second line in Table 2.1.

For the plaintext: I love you, the correspondent cyphertext would be "osgctngx."

The first documented use of a substitution cipher was by Julius Caesar, during the famous Gallic Wars. The most famous system used by Caesar was called the "Caesar Cipher." There the sender of the message replaces each letter of the plaintext by the letter that is three places further down in the alphabet. For example, the letter A would be substituted by D, B by E, and so on. It is clear that this system is not secure. The total number of possible keys is 25; therefore, an exhaustive key search is possible even without a computer. Just try every possible key K until a meaningful plaintext string is obtained. We observe that a necessary condition for a cryptosystem to be secure is a large set of possible keys, but this condition is not enough to ensure security. For example, a general substitution technique provides a key space with 2^{26} possible keys. However, a way to break a

Table 2.1
An Example of a Substitute Table

Plaintext	A	B	C	D	E	F	G	H	I	J	K	L	M
Ciphertext	Q	W	E	R	T	Y	U	I	O	P	A	S	D
Plaintext	N	O	P	Q	R	S	T	U	V	X	W	Y	Z
Ciphertext	F	G	H	J	K	L	Z	X	C	V	B	N	M

general substitution cipher is to use the so-called frequency analysis. In frequency analysis, the frequency of appearance of each character in the ciphertext is computed and compared with the frequency of appearance of characters in a normal text written in the language used to write the message. Closeness between frequencies of characters in the plaintext and in the ciphertext indicates potential encryption/decryption rules.

Another famous example of a substitution cipher is the Enigma, the enciphering system used by the Nazis in Germany during the Second World War. A big innovation of the Enigma is that it was the implementation of a cryptographic algorithm in a machine and ended the pencil-and-paper era of cryptography. Enigma was an astonishing advance in several aspects, making possible the encryption of a large amount of documents in very little time. It also allowed more complicated (and more secure) algorithms to be implemented, making it a formidable and intricate enciphering machine.

The Enigma machine began life in 1923 as a commercial product produced by a German named Arthur Scherbius, and aimed at businesses with a need for secure communication. It was a simple device to use. After setting it up, the operator types in the plaintext of his message. Each time a key is pressed, a letter on the lampboard is illuminated corresponding to the ciphertext. The operator then simply notes down the ciphertext letter and carries on.

The Germans perceived the Enigma as being practically unbreakable. They were wrong. The Enigma and the procedures with which it was used had a number of weaknesses that were exploited by the Allies to probe into the tactical and strategic secrets of the German military.

2.5.3 Diffusion/Confusion

In his classical work, Shannon stated that good cryptosystems should provide confusion and diffusion. Confusion means that the relationship between plaintext and ciphertext should be as complicated as possible. Diffusion means that the influence of each bit of the plaintext over the ciphertext should be as great as possible, thus yielding ciphertext statistical properties apparently unrelated to the plaintext.

Shannon also proposed to combine weak and maybe insecure simple ciphers to obtain stronger and potentially secure cryptosystems with good confusion and diffusion properties. These ciphers, made from the combination of simpler ones, are called product ciphers. A trivial way of implementing a product cipher is to encrypt a message twice, once using a given key and then to encrypt the ciphertext resulting from the first encryption by using a different key. The decryption is done in a similar way, but in the reverse order.

2.5.4 SP Networks

The concept of product ciphers is also used in the so-called SPnetworks, where "S" stands for substitution and "P" for permutation. SP networks constitute the most basic building block in a wide range of available block ciphers. In an SP network, the message is substituted and permuted repeatedly in rounds. The substitutions are performed by algorithms called S-boxes, lookup tables that map n bits to m bits (where n and m often are equal). Usually, bits from the key are used to determine which substitution is applied to the plaintext. S-boxes increase the confusion of the ciphertext. The permutations are common tools in mixing bits (increasing the diffusion). They are linear operations, and thus not sufficient to guarantee security. However, when used with good nonlinear S-boxes they are vital for the security because they propagate the nonlinearity uniformly over all bits. These concepts are illustrated in Figure 2.3.

Several modern ciphers have their structures based on SP networks.

2.5.5 Basic Structure

Most of the currently used block ciphers are based on SP networks. Of particular interest is an SP network known as Feistel network, which is composed of several rounds. In each round, a block of data is used as input and an output block of the same size is generated. All of these rounds basically repeat the same operations. Initially, the message is split into two halves. The second half is used as the input to a nonlinear function F (which usually depends on the secret key used in the encryption process). The output of this nonlinear function is then XORed with the first half of the input message. The

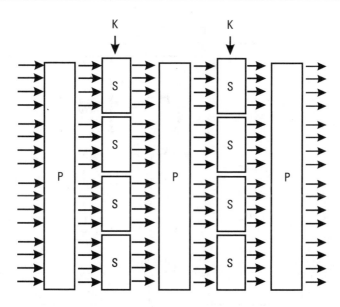

Figure 2.3 SP networks.

output of the XOR operation will be the second half of the output block. The first half of the output block is a repetition of the second half of the input block. These procedures are illustrated in Figure 2.4.

The function F is used to increase the confusion of the ciphertext, while the swapping (inversion of the first and second halves) is used to increase the diffusion.

2.5.6 Modes of Use

There are several ways one can encrypt data by using a block cipher. The most popular are:

- Electronic codebook book (ECB): In this method, the message is broken into independent blocks with length equal to the length used by the block cipher, and then encrypted with the same key.

- Cipher block chaining (CBC) (see Figure 2.5): Here again, the message is broken into blocks, but linked together in the encryption operation with an initialization vector. An initialization vector with length equal to the length used by the

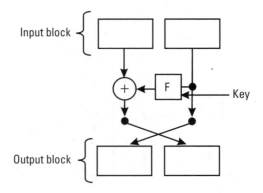

Figure 2.4 Feistel networks.

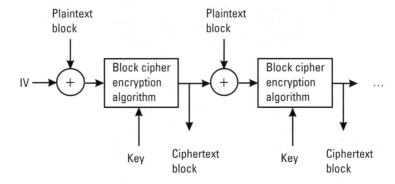

Figure 2.5 The cipher block chaining mode.

block cipher is then encrypted. This encrypted block is XORed with the next plaintext block and the result is then encrypted again yielding a new ciphertext block. The process is repeated until all the plaintext blocks have been encrypted.

- Cipher feedback (CFB) (see Figure 2.6):The message is treated as a stream of bits, added to a stream of bits generated by the block cipher from an initialization vector. The encryption proceeds as follows: the initialization vector is encrypted and the result is XORed with the first block of plaintext generating the first ciphertext block. This ciphertext block is then encrypted again and the output is XORed with the second plaintext block with the result being feed back for the next stage.

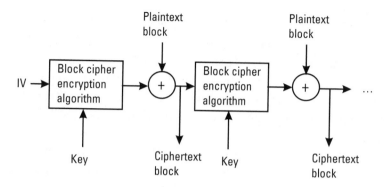

Figure 2.6 The cipher feedback mode.

- Output feedback (OFB): In this case, the message is treated as a stream of bits, as in the CFB mode, but with the feedback being independent of the message. An initialization value is then both encrypted and XORed with the first plaintext block. To generate the second ciphertext block, the encrypted initialization value is then encrypted again and the result is XORed with the second plaintext block. The process is repeated till all the plaintext blocks are encrypted.

ECB mode is less secure than the other modes, since identical plaintext blocks are encrypted into the same ciphertext blocks. CBC and CFB modes are more secure, however. If there are errors made during the transmission of the ciphertext, these errors are propagate, through the decryption algorithm, to all the subsequent blocks. Therefore, in spite of its weakness, ECB is preferred when data transmission errors are too frequent, such as in satellite transmissions.

Although these are general modes of use that apply to any block cipher, they were designed mostly for use with DES. New modes of use are in development process for the new AES.

2.5.7 DES

Arguably, the most famous block cipher is the Digital Encryption Standard (DES). DES was the first algorithm to be proposed by an official agency of a government, in this case, the National Security Agency (NSA) of the government of the United States.

DES is a block cipher with 64-bit block size. It uses 56-bit keys. The same algorithm is used with the same key to convert ciphertext back to plaintext. The DES consists of 16 "rounds" of operations that mix the data and key together in a prescribed manner using the fundamental operations of permutation and substitution. Most of the currently used block ciphers are based on SP networks and DES is no exception. Actually, DES structure can be viewed as a Feistel network. The goal is to completely scramble the data and key so that every bit of the ciphertext depends on every bit of the data plus every bit of the key (a 56-bit quantity for DES). DES is susceptible to exhaustive key search with modern computers and special-purpose hardware. DES is still strong enough to avoid most random hackers and individuals, but it is easily breakable with special hardware by government, criminal organizations, or major corporations.

A variant of DES, triple-DES (also 3DES) is based on using DES three times. Triple-DES is arguably much stronger than (single) DES, however, it is rather slow compared to some new block ciphers.

2.5.8   AES

The Advanced Encryption Standard (AES) was the name chosen for the substitute of DES. It was selected among several candidates by the National Institute of Standards and Technology (NIST). The chosen cipher was one proposed by Belgium researchers, and called Rijndael, named after its inventors, Rinjmen and Daemen.

Rinjdael is a block cipher with block length 128 and three possible key sizes, 128 bits, 192 bits, and 256 bits (parameters specified by NIST). Although it is based on an SP network architecture, Rijndael is not based on Feistel networks. The number of rounds depends on the key length; it is 10 if the key length is 128 bits, 12 if the key lenght is 192 bits, and 14 if the key length is 256 bits.

Rijndael is apparently resistant against all the known attacks against block ciphers. At this time, the best attack against it is exhaustive key search.

The most common critic to AES is that it can be expressed in a transparent algebraic way. It can be show that to break AES is equivalent to solve a certain class systems of polynomial equations over a defined finite field. Even though there is currently no efficient algorithm known for solving those systems of polynomial equations, some researchers in the community fear that AES' clean mathematical structure may be open to an attacker in future.

2.6 Stream Ciphers

Stream ciphers are symmetric key encrypting algorithms that work on single bits of the plaintext (Figure 2.7). While they are similar to the one-time pad, instead of XORing the message to a pure random string, stream ciphers XOR the message to a pseudorandom sequence of bits, usually called a keystream, generated from a small sequence of pure random bits. A sequence is pseudorandom if is computationally infeasible for an adversary to distinguish between the sequence and a real random string. The decryption proceeds by XORing the ciphertext with a keystring generated from the same small sequence of pure random bits. Block ciphers can also be used as stream ciphers when working in the CFB or OFB modes.

Stream ciphers are usually categorized as synchronous or self-synchronizing.

- Synchronous stream ciphers: the generation of the keystream can be independent of the plaintext and ciphertext.

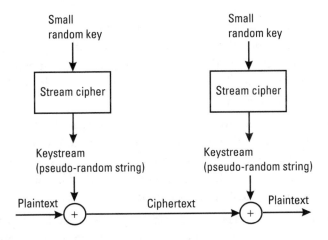

Figure 2.7 Stream cipher.

- Self-synchronizing stream ciphers: the generation of the keystream depends on the data and its encryption.

2.6.1   Advantages

There are several advantages to using this approach over the one-time pad and over block ciphers, namely:

- Different from the one-time pad, the size of the key can be much smaller than the size of the message to be encrypted.
- The encryption operation is very fast; usually stream ciphers are much faster than block ciphers.
- There is no error propagation, that is, if parts of the cipher text are corrupted during the transmission, it does not affect other uncorrupted parts.

Note that stream ciphers are not based on the confusion/diffusion principle created by Shannon.

2.6.2   Security Remarks

Following are some points to be understood when studying stream ciphers:

- If the key is used more than once, the system can be compromised, since if an adversary has two ciphertexts encrypted with the same key, he can add up the two ciphertexts and obtain the XOR of two messages, which usually can be separated very easily.
- Another point that should be noted is that all stream ciphers are periodic; that is, they start repeating the keystream output after a while. Therefore, if very long messages are encrypted, it has the same effect of using a key twice. The period of repetition depends on the design of the stream cipher.
- Finally, we have to consider the single-cycle property. A stream cipher has the single-cycle property if there is only one periodic sequence of numbers it generates. Thus, keys are

only shifted windows of this unique periodic sequence. Of course, the periodic sequence may be very large, but it is still better to avoid stream cipher with this property.

2.6.3 Some Examples

Following are some examples of proposed stream ciphers:

- *RC4:* an algorithm proposed by Ron Rivest (one of the creators of RSA). It is very simple even tough some recent attacks were proposed (weak keys and attacks which work when RC4 is used for broadcasting data) it remains secure for most of the applications if properly used. Its key size is 40 bits.

- *SCREAM:* SCREAM (designed by Halevi, Coppersmith, and Jutla) is an improvement over a previous stream cipher called SEAL, designed by Phil Rogaway and Don Coppersmith. It was optimized for achieving very fast software implementations. No practical attacks are known against it.

- *MUGI:* A stream cipher designed by Hitachi. Its design is based on a previously proposed cipher known as Panama. The authors claim it yields efficient hardware and software implementations.

- *Snow 2.0:* A word oriented stream cipher (thus allowing a faster implementation in software). Its word size is 32 bits and two key sizes can used with it: 128 or 256 bits. Together with MUGI, Snow 2.0 has been suggested as possible standards for ISO.

2.7 Asymmetric Cryptosystems and Digital Signatures

We now turn our attention to asymmetric cryptosystems. We review basic concepts of public key encryption, digital signatures, the necessary mathematical background, and the adversarial model.

2.7.1 Public-Key Encryption

Confidentiality of transmitted data cannot be guaranteed by the sole use of symmetric schemes. Users must share a common key for the utilization of a symmetric key encryption scheme. If the users cannot securely carry out their key establishment, an eavesdropper may illegally obtain the key, hence, revealing the transmitted message. One of the most practical solutions to this problem is to use public key encryption in which a sender encrypts a message (or a session key) by using only the receiver's public information.

The idea of public-key encryption (using asymmetric algorithms) was proposed by Diffie and Hellman in their pioneering paper in 1976, where the keys for encryption and decryption were called public and private key, respectively. Their revolutionary idea was to enable secure message exchange between sender and receiver, without ever having to meet in advance to agree on a common secret key. Public key encryption can be (roughly) modeled as follows:

In a situation where Bob intends to send a message to Alice, she first generates a public key which is published under her name in a public directory accessible for everyone to read and a private-key, which is known only to her. To send a secret message to Alice, Bob looks up Alice's public key from the public directory, and then encrypts the message using the public key. The resulting ciphertext (the encrypted message) will be sent to Alice via a public channel, such as the Internet. Finally, upon receiving the ciphertext, Alice can decrypt and recover the message by using her private key (see Figure 2.8). The public key encryption mechanism is based on the difficulty of solving certain types of mathematical problems (e.g., integer factoring and discrete logarithm), and this ensures it would take an inordinate amount of time to figure out the private key from the public one.

Currently, there are many public-key encryption techniques that have been intensively developed and which provide different levels of security and efficiency. One of the most important developments is on the topic related to provable security. This is the security used to rigorously prove the impossibility of an adversary to break a cryptosystem unless he is able to solve a mathematically hard problem posed to him. Before, heuristic approaches were often used to analyze the security of cryptographic schemes (i.e., if no one was able to "break" the scheme after many years, then its security was widely

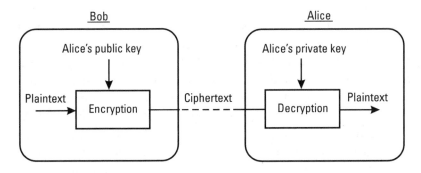

Figure 2.8 A public key cryptosystem.

accepted. However, that approach is shifting towards provable security. Provable security is becoming an indispensable requirement for a standard cryptosystem, and hence, most of the current cryptosystems fulfill this requirement. Later in this chapter, we analyze provable security in details.

Needless to say, one of the main advantages of public key encryption is the establishment of secure communication by using only public discussions. In addition to this, the number of keys corresponding to each communication partner that a single user needs to manage in conventional schemes is significantly reduced in public key encryption, where a user has to only keep an eye on her private key.

At first glance, it seems likely that in all aspects, public-key encryption is superior to symmetric key encryption. However, this is not always true. One of drawbacks of (existing) public key encryption schemes is speed: They require heavy computational costs and their message size is significantly limited. Nevertheless, the best solution may be to combine both the advantages of conventional and public-key encryption. Namely, it may not be practical to encrypt the whole message by using only public key encryption, and it may be wiser to use public key encryption to encrypt a session key instead, which will then be utilized as a common key between a sender and a receiver.

2.7.2 Public Key Infrastructure (PKI) and Certificate Authorities (CA)

While one might think public key encryption provides sufficient security, there are disadvantages. Even if a user's private keys are not available, a successful attack of the public database will allow an adversary to impersonate the user by substituting his public key. It is therefore important to install a mechanism that guarantees the relationship between the public key and the identified person. A practical and typical solution to this problem is to enlist a third party trusted by both the sender and recipient known as certificate authority (CA). CA ascertains the identity of a user and certifies that the public key assigned to him/her in the public database actually belongs to the user. Once the CA verifies the association between an identified user and his public key, CA issues a certificate. A certificate is a computer-based record that attests to the binding of a public key to an individual or entity. If the recipient wants to verify the connection between the sender and his public key, the recipient can look to the attached certificate. CA must be an objective third party with an established reputation for credibility to give reliable credibility to the issued certificates (see Figure 2.9.)

A certificate contains the name of the sender, his or her public key, a serial number, expiration dates, and, most importantly, the digital signatures of the CA enabling a recipient to verify that the authenticity of the certificate (Figure 2.10).

In a large scaled network, authorization of public keys can be considerably burdensome, and a single CA insufficient for issuing certificates and verifying the identities of all users. Although a solution may seem to be setting up multiple CAs, it is difficult to concurrently manage many (fully trusted) CAs. Typically, a hierarchical network of CAs is constructed: a root CA serves to sublicense parent CAs by issuing certificates who will in the same manner authorize their children CAs. Each additional CAs will generate certificates for

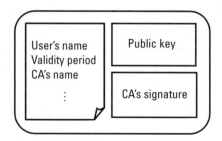

Figure 2.9 A certificate issued by a certificate authority.

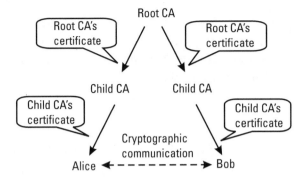

Figure 2.10 Hierarchical CAs.

a small subset of users who may serve as the leaf node of the entire hierarchy. These functions provide minimal interruption for certificate authentication services.

2.7.3 Mathematical Background

Before continuing to delve further into the details of public key encryption, a brief review of some underlying mathematics is needed.

Suppose Bob sends a message to Alice by using public key encryption. Let the message, its corresponding ciphertext, the recipient's private key and his public key be m, c, $sk_{Alice}(x)$, and pk_{Alice}, respectively. Since c is calculated using m and pk_{Alice}, c can be mathematically expressed as $c = f_{pkAlice}(m)$, where $f_{pkAlice}(x)$ is an encryption function depending on pk_{Alice}. We note that in a practical public key encryption system, a random number is also chosen by Bob, and this will be used for generating c as well. A necessary (but not sufficient) condition to carry on "secure" public encryption is to let the recipient's private key, $f_{pkAlice}(x)$ have the following property: it must be significantly hard to calculate x_0 from $f_{pkAlice}(x_0)$ for all x_0. You may want to think of it as a process that is easy to accomplish, but very difficult, or even impossible, to undo. The function satisfying this condition is called a one-way function. (Alice is able to "break" the one-wayness by using her private key $sk_{Alice}(x)$ more precisely, $f_{pkAlice}(x)$ is required to be a trapdoor one-way function.)

All practical public-key encryption systems are based on (trapdoor) one-way functions. The following is an example of such one-way function commonly used in public key cryptography:

Recall that for integers a, b and c, $a \bmod b = c$ means that dividing a by b, we get the reminder c. Define $f_{g,p}(x) = gx \bmod p$. A function $f_{7,11}(x)$ is expressed as $f_{7,11}(x) = 7^x \bmod 11$ and for $x = 1, 2, \ldots, 10$ it is efficiently calculated as follows:

$$f_{7,11}(1) = 7 \bmod 11 = 7$$

$$f_{7,11}(2) = f_{7,11}(1) * f_{7,11}(1) \bmod 11 = 5$$

$$f_{7,11}(3) = f_{7,11}(2) * 7 \bmod 11 = 2$$

$$f_{7,11}(4) = f_{7,11}(2) * f_{7,11}(2) \bmod 11 = 3$$

$$f_{7,11}(5) = f_{7,11}(2) * f_{7,11}(3) \bmod 11 = 10$$

$$f_{7,11}(6) = f_{7,11}(3) * f_{7,11}(3) \bmod 11 = 4$$

$$f_{7,11}(7) = f_{7,11}(3) * f_{7,11}(3) * 7 \bmod 11 = 6$$

$$f_{7,11}(8) = f_{7,11}(2) * f_{7,11}(3) * f_{7,11}(3) \bmod 11 = 9$$

$$f_{7,11}(9) = f_{7,11}(3) * f_{7,11}(3) * f_{7,11}(3) \bmod 11 = 8$$

$$f_{7,11}(10) = f_{7,11}(3) * f_{7,11}(3) * f_{7,11}(3) * 7 \bmod 11 = 1$$

However, for the next problem: for a given y_0, calculate x_0 such that $y_0 = f_{7,11}(x_0)$, you can see that it is much more difficult to obtain the answer. In fact, there is no known efficient method to solve this problem. Immediately, we can calculate $f_{7,11}(4)$ to be 3, but it is not easy to compute x_0 where $f_{7,11}(x_0) = 3$ (Maybe, for the above problem, the fastest way may be to produce a table of possible pairs of x and $f_{7,11}(x)$ for all x).

The discrete logarithm problem is the problem of finding the inverse of $f_{g,p}(x)$ and it plays an important role in many cryptographic schemes. The difference in the computational cost for $f_{g,p}(x)$ and that for its inverse increases significantly as the parameters increase. For large parameter settings, we can assume that function $f_{g,p}(x)$ is a one-way function. Roughly speaking, computation of an inverse of

$f_{g,p}(x)$ is considered infeasible when 1,024-bit-long p (or larger) is selected (while other parameters are appropriately chosen). Because of its intractable property, we are able to construct a trapdoor one-way function which is the heart of public key encryption.

Like the discrete logarithm problem, integer factoring problem is considered to be a hard problem as well: letting p and q be large primes, and n be $p*q$, then, for a given n, finding p and q is considered be difficult. This problem becomes extremely difficult when the numbers get larger. RSA cryptosystems is based on the integer factoring problem which implies that RSA can be broken if you can solve the underlying integer factorization.

2.7.4 Diffie-Hellman Key Agreement

The Diffie-Hellman (DH) key agreement [3] provided the first solution to the key distribution problem, allowing two parties, never having met in advance, to share a secret key by public discussion only. Key agreement protocols are different from public-key encryption protocols in that they are not meant for encrypting communications, but rather, they are used to agree upon a secret that can be used for encrypting communications. So far, many secure and practical key-exchange algorithms have been developed and the Diffie-Hellman key agreement technique is definitely one of the classic proposals and still used in many applications.

Suppose that Alice and Bob want to agree on a shared secret key using the Diffie-Hellman key agreement protocol. First, Alice generates her private keys by picking a random value, x_{Alice} between 1 and p-2, where p is a large prime number. Similarly, Bob generates his private key x_{Bob}. Then, for an appropriate public parameter g (i.e., g is a "generator" of a cyclic group of order p-1 in $\{1,2,...,p$-1$\}$), Alice and Bob derive their public keys, $y_{Alice} = f_{g,p}(x_{Alice})$ and $y_{Bob} = f_{g,p}(x_{Bob})$, respectively. They proceed by exchanging their public keys. Finally, Alice computes $K = fy_{Bob,p}(x_{Alice})$ and Bob computes $K' = fy_{Alice,p}(x_{Bob})$ Since $K = K'$, Alice and Bob can use this key as their secure secret key to encrypt their communications for this session.

Currently the best-known method to break the Diffie-Hellman key agreement protocol is to solve the underlying discrete logarithm problem, which is believed to be one-way.

2.7.5 RSA Cryptosystem

The Rivest-Shamir-Adleman (RSA) cryptosystem [4] is one of the most widely accepted public key encryption methods. The RSA Algorithm is based on the difficulty of factorizing large numbers.

RSA uses modular arithmetic and elementary number theory to do certain computations. It proceeds as follows: Alice chooses two large primes p and q and computes $n = p{*}q$. Alice further calculates l which is the least common multiple of p-1 and q-1. She then selects a random integer e such that $1 < e < l$ where e and l are mutual primes. Alice also computes d such that $e{*}d = 1$ mod l (this procedure can be efficiently performed by well-known mathematical techniques). Alice's public key is the pair: n and e, and Alice's private key is d.

To encrypt a message m, where m is no larger than n-1, Bob (the sender) calculates the corresponding ciphertext c as follows:

$$c = m^e \bmod n$$

and Bob sends c to Alice.

To decrypt c, Alice computes

$$c^d \bmod n = m$$

The security of the RSA cryptosystem is based on the intractability of the integer factoring problem [it is believed that it is extremely difficult to factor very large numbers (100–200 digits)]. If an adversary is able to factor n into p and q and compute the decryption key, d, his attack is considered successful. Discovering d would cause the worst possible damage (i.e., an adversary will be able to read all messages encrypted with the public key and forge signatures as well). Since no other methods for determining d without factoring n are known, a large enough n should be used to enhance the security of RSA. Currently, 1,024-bit long n is recommended for use in practical systems.

Next, we consider a stronger notion of security of RSA. "Textbook" RSA, which is the straightforward implementation of the above description, is considered to be insecure in a restrictive sense. This is primarily due to the following reasons: even if it is impossible to break the one-wayness of RSA encryption function, it is still possible to obtain certain plaintext information from the ciphertext.

As an extreme example, consider the case where an adversary knows that Bob is going to send a simple yes or no message to Alice. An adversary can easily eavesdrop the message by obtaining Bob's transmitted ciphertext and comparing it with the ciphertext generated by encrypting the word yes with Alice's public key. If these two ciphertext are identical, the adversary, without ever having to decrypt it, would know that the transmitted message was yes, and if different, it was no. In another situation an adversary may be allowed to create any ciphertext (except for the target ciphertext) and ask Alice to decrypt it and get to see the corresponding plaintext at his will. If this is possible, the adversary can obtain the plaintext of the target ciphertext (without having to ask Alice to decrypt the target ciphertext). This type of attack model is called chosen ciphertext attack.

Initially, developers did not consider protecting their systems against such a strong attack environment. However, by the end of 1990s, it was shown by Daniel Bleichenbacher of Bell Labs, a division of Lucent Technologies, that an adversary is in fact capable of performing such kind of attacks on practical cryptosystems. He found a flaw in certain encryption protocol based on RSA, the dominant standard for encrypting and decrypting data used widely. Protection against attacks as chosen ciphertext attack has now taken on greater importance and new cryptosystems are currently implemented in a more secure way. RSA cryptosystems, for example, are now generally padded with optimal asymmetric encryption padding (OAEP) [5] which is a type of digital envelop format, a well-known method for enhancing its security.

2.7.6 ElGamal Cryptosystem

The ElGamal Cryptosystem [6] is a public key encryption scheme based on the discrete logarithm problem. The Diffie-Hellman key agreement, too, is based on the intractability of discrete logarithm problem. However, restrictively speaking, it is not a public key encryption. On the other hand, ElGamal cryptosystem is a modified version of the Diffie-Hellman key agreement and also provides the functionality of public key encryption.

The encryption algorithm of the ElGamal cryptosystem is similar in nature to the Diffie-Hellman key agreement. Consider that Bob sends a message to Alice. First, Alice picks her secret key, x (no larger than p-2, where p is a large prime) uniformly at random. Also, for an appropriate g, Alice calculates her public key, $y = f_{g,p}(x)$ and publicizes it.

To encrypt a message m (again, m is no larger than $p - 1$), Bob first looks up Alice's public key $y = f_{g,p}(x)$. Then he chooses a random "one-time'" key, r (no larger than $p - 2$), and calculates the corresponding ciphertext $\{c_1, c_2\}$ as follows:

$$c_1 = g^r \bmod p$$

$$c_2 = m * y^r \bmod p$$

and Bob sends $\{c_1, c_2\}$ to Alice.

To decrypt $\{c_1, c_2\}$ Alice computes

$$c_2 * \left(c_1^x\right)^{-1} \bmod p = m$$

where $(c_1^x)^{-1}$ is defined as $(c_1^x)^{-1} * c_1^x \bmod p = 1$

Unlike the RSA cryptosystem, which involved two keys, one public and one secret, the ElGamal cryptosystem uses a random number as a third key for encryption. It is sort of a "disposable" key, used just once to encrypt just one message, and there may exist multiple ciphertexts derived from the same message (this type of cryptosystem is called probabilistic encryption). While essential to the functioning of the ElGamal cryptosystem, it does not appear in other public-key cryptosystems and does not play a role in the concept of a public-key cryptosystem. However, it guarantees that any observation of the ciphertext does not provide any meaningful information on the corresponding plaintext. For example, if the possible plaintexts are the only two, yes and no, for a given ciphertext, an adversary can try to guess what the plaintext is by comparing the ciphertext with another ciphertext which he generates by encrypting yes. It is unlikely, however, for these two ciphertexts (even if these are produced from the same message), to be identical. Therefore, seeing different encryptions of the same message does not help an adversary to figure out the original message.

However, textbook ElGamal, like textbook RSA, is not secure against chosen ciphertext attacks. To be more specific, when Bob sends an encrypted message, $\{c_1, c_2\}$ to Alice, an adversary can illegally decrypt the ciphertext as follows: the adversary chooses a random number, r^1 and asks an oracle (this oracle is present in some security models such as chosen ciphertext security) to decrypt $\{c_1, r^1 * c_2\}$. The oracle's answer will be $r^1 * m$, and so the adversary obtains m by a simple division, $r^1 * m/r$. It is therefore important to take such problem into account (considering whether it is worth achieving such a high level of security as chosen ciphertext security) when implementing cryptosystems. Some methods that can enhance the security of standard ElGamal cryptosystem are introduced in the following sections.

2.7.7 Necessary Security for Practical Public Key Encryption Systems

As discussed earlier, standard (or textbook) RSA and ElGamal cryptosystems are vulnerable to certain type of attacks, and straightforward implementation of these schemes may not be suitable in practical systems. Therefore, when constructing a practical communication system based on the standard RSA or the ElGamal cryptosystem, it is necessary to enhance these systems by some means to ensure a certain security level.

Currently, it is desirable to satisfy the following security requirement when implementing practical public key encryption systems: *For any ciphertext, any adversary cannot obtain any meaningful information on a plaintext even if he or she is allowed to ask the receiver to decrypt any ciphertext excluding the target ciphertext* [7,8]. This means that in a practical, public key encryption system, it must prevent any partial break even against the most advantageous adversary.

For many years, such a strong notion of security in public key encryption had been thought of as a theoretical issue only, and the majority of developers at that time did not feel necessary to eliminate and control threats and vulnerabilities as of such an elevated level when building cryptosystems. However, in the late 1990s, it was identified by Daniel Bleichenbacher of Bell Labs that in a practical situation, RSA cryptosystem in PKCS#1ver1.5 (which gained remarkable popularity and was the most widely used protocol) was in fact vulnerable against chosen ciphertext attacks [9]. This incident had brought many developers to realize the need of protecting their systems from the subjected attack, and when implementing, counter-measures should be taken to thwart the attack.

OAEP is a well-known method for encoding messages used to enhance cryptosystems such as RSA, to provide provable security against chosen ciphertext attacks. OAEP can be thought of as a pre-processing performed before encrypting it with RSA, which converts the plaintext to a form that can be enciphered by RSA encryption. The improved RSA cryptosystem with OAEP is provably secure; that is, it is secure in the sense that no adversary is able to obtain any meaningful information on the plaintext even if he is able to perform a chosen ciphertext attack. Currently, OAEP is widely used in the implementation of the RSA cryptosystem in PKCS#1ver2.1.

For the ElGamal cryptosystem, while it is difficult to apply OAEP in a straightforward manner; other methods to enhance the security of ElGamal cryptosystem have been proposed [10,11]. By applying these techniques, it is also possible to achieve provable security against chosen ciphertext attacks for the ElGamal cryptosystem.

2.7.8   Digital Signature

A digital signature is an electronic substitute for a handwritten signature. It is an identifier created by a computer instead of a pen. Similar to our physical communication systems, it is also important to be able to securely "sign" a message to prohibit substitution or impersonation by dishonest users for electronic communications systems. Various methods for signing a digital data have been developed for practical use and are utilized in a number of applications, such as e-commerce.

So, now, how do we digitally sign a message? Before Alice can digitally sign an electronic communication she intends to have with Bob, she must first create a public-private key pair. The private (signing) key is kept confidential, and is used for the purpose of creating digital signatures. As the public (verifying) key is disclosed, the recipient of the digitally signed message can access it. Alice then selects a message m to be signed. Her digital signature $Sig_{Alice,m}$ is generated as a function of her private key and m. On receiving the digitally signed message, m and $Sig_{Alice,m}$, Bob can verify the validity of the signature by using Alice's public key. The verification of the signature is performed as a function of m, $Sig_{Alice,m}$ and Alice's public key. Moreover, unlike a handwritten signature, which is unique to the signer, but presumably consistent across all documents signed, a digital signature is unique for each document signed. This is because a digital signature is derived from the document to be digitally signed. Any change to the document will produce a different digital signature (Figure 2.11).

Any user is able to verify digital signatures since no secret information is required for verification. When a recipient obtains the public key of someone from whom he has received a digitally signed message, he must determine for certain that the public key does, in fact, belong to the purported sender. Thus far, the discussion has made the one critical assumption that the public-private key pair of the sender does, in fact, belong to the sender. In order to prevent substitution attack of a public key, the relationship between the public key (public-private key pair) with the identified person must be guaranteed. The solution to this problem is to enlist a third party trusted by both the sender and the recipient, or, a certificate authority (CA). CA certifies that the public key of a public-private key pair used to create digital signatures does belong to that person.

2.7.9 Mathematical Background

Just like in public key encryption, one-way functions also play an important role in digital signature mechanism. This fact can be immediately verified from the following observation:

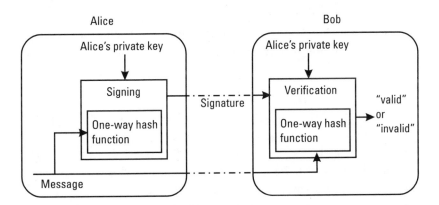

Figure 2.11 A general digital signature scheme.

For a fixed message *m*, signature generation can be regarded as a function of a signer that is Alice's private key. Therefore, it is not possible to recover Alice's private key using the signature itself and other public data only. It follows that the signing function is also a one-way function, and hence, it is impossible to construct a digital signature system without one-way functions. To be precise, it was proven that from any one-way function, a digital signature digital signature scheme can be constructed. Namely, existence of a one-way function is a necessary and sufficient condition for the existence of a secure digital signature scheme.

There are primarily two types of one-way functions generally used in digital signature systems. One is an algebraic one-way function that provides the main functionality of generating signatures, and these functions are based on the discrete logarithm problem or the integer factoring problem. The other is called (cryptographic) one-way hash functions that ensure impossibility of substitution of a publicized signature. Here, we focus on the latter, or one-way hash functions. Roughly, a one-way hash function is required to satisfy the following properties:

- Any message with an arbitrary length can be an input.
- The size of an output is fixed, and is a constant value.
- It is relatively easy to compute the output from a given input.
- For a given output of a function, it is difficult to find the corresponding input, or the function is one-way.
- There should not be any different inputs such that their corresponding outputs are identical, or the function is collision-free.

In most digital signature schemes, it is typical to begin a signing transaction by generating a message digest using a one-way hash function. Instead of computing the signature on the message itself, message digest (or the hashed value of the message) is used to derive the signatures. This is due to the fact that it is much easier to compute signatures from message digests rather than more lengthy messages.

2.7.10 RSA Signature Scheme

The RSA cryptosystem can be easily converted into a digital signature scheme. The RSA signature scheme is widely used in many practical systems, as, for example, PKCS#1. In an RSA digital signature scheme, Alice chooses two large primes p and q, computes $n = p*q$ and calculates l, which is the least common multiple of $p-1$ and $q-1$. Then, Alice selects a random integer e such that $1 < e < l$ where e and l are mutual primes. She also computes d such that $e*d = 1 \bmod l$ (this procedure can be efficiently performed by well-known mathematical techniques). Alice's public key is the pair, n and e, and her private key is d.

To sign a message m, Alice calculates her signature σ as follows:

$$\sigma = H(m)^d \bmod n$$

where H is an appropriate one-way hash function. m, σ will then be published as Alice's signed message.

Receiver Bob accepts m, σ if the following equation holds:

$$\sigma^e \bmod n = H(m)$$

Intuitively, using the above testing, Bob can confirm that σ is equivalent to $H(m)^d \bmod n$ since $H(m)^{d*e} = H(m) \bmod n$, and this fact implies that σ has actually been generated by Alice since she is the only one who knows d. As you can see, the use of the one-way hash function is crucial in this case. Without a one-way hash function, an adversary will be able to easily forge Alice's signature: $(m_1 \sigma_1)$ and $(m_2 \sigma_2)$ be Alice's valid pairs of a message and a signature actually generated by her. By using them, an adversary will be able to calculate $\sigma_3 = \sigma_1 * \sigma_2$, which is Alice's valid signature derived for message $m_3 = m_1 * m_2$. It should be clear now that it is indispensable to utilize one-way hash function when implementing RSA signatures to prevent such types of attacks. However, this does not imply that RSA signature scheme is secure if a one-way hash function is used for generating message digest.

In fact, in 1999, a practical implementation of the RSA signature, standardized as ISO9796-2, was shown to be vulnerable. Since then, provable security against all possible attacks (and not just against certain attacks) is desired even for practical systems, and several methods for enhancing the security of the RSA signature scheme have been proposed, see Section 2.7.12.

2.7.11 Digital Signature Algorithm (DSA)

Digital Signature Algorithm (DSA) is a digital signature scheme proposed by the U.S. National Institute of Standards and Technology (NIST) as a U.S. Federal Information Processing Standard (FIPS 186). Security of DSA is based on the discrete logarithm problem.

In DSA, user Alice picks primes p and q such that q divides (p-1), and sizes of p and q are 1,024 and 160 bit, respectively. Let g be an appropriate element from $1,2,...,p$. (More specifically, g is a "generator" of a cyclic group of order q in $\{1,2,...,p\}$) Then, Alice chooses an integer x (a positive integer smaller than $q-1$) as her private key and computes $y = g^x \bmod p$. Alice's public key is (p,q,g,y).

When signing a message m, Alice first calculates $H(m)$ where H is a secure hash algorithm (SHA-1), which is a one-way hash function. Then, Alice selects a random secret integer k (a positive integer no larger than $q-1$), and computes

$$r = \left(g^k \bmod p \right) \bmod q$$

$$s = k^{-1} \left(H(m') + xr \right)$$

where k^{-1} is an integer such that $k^{-1} * k = 1 \bmod q$. The pair (r,s) will be published as Alice's signature for m.

Receiver Bob accepts (m,r,s) if the following equation holds:

$$r = g^{s^{-1}*H(m)} y^{r*s^{-1}} \bmod p \bmod q$$

where s^{-1} is an integer such that $s^{-1} * s = 1 \bmod q$.

2.7.12 Necessary Security for Practical Digital Signature Systems

As described so far, there are many security issues that need to be considered to establish secure and practical digital signature systems. It was shown that a straightforward implementation of RSA signature is vulnerable to certain attacks, and moreover, it has still not been established whether or not DSA satisfies all the security requirements of a digital signature scheme. Recall that currently, for a digital signature system used in practical system, it is desired to satisfy the following security requirement: *for any message, any adversary cannot forge a valid signature that the valid signer has not generated even if the adversary is allowed to ask the signer to sign any message at his will* [12]. This means that a practical signature system must prevent any existential forgery even against the most advantageous adversary.

Similar to the security discussion given for public key encryption schemes earlier, we can say about the same for digital signature schemes. The nature and mechanism of the threat against digital signature schemes was poorly understood, and the majority of developers did not assess their products to a newly discovered potential security attack and did not consider taking effective countermeasures to thwart such attacks. However, by the end of the 1990s, it was shown that an implementation of the RSA signature scheme, which was standardized as ISO9796-2, was also vulnerable against chosen message attacks [13].

There are two primary methods to protect and enhance the RSA signature by providing provable security for the above security notion. One is called full domain hash scheme (FDH)[14], and the other, probabilistic signature scheme (PSS)[14].

Since RSA signature scheme was proven to be vulnerable to certain attacks if the message digest is small enough to be factorized, possible countermeasures can be taken, such as increasing the size of the hashed value (by one-way hash function) to be large as possible. The RSA signature scheme whose message is hashed onto the full domain of the function is, hence, called full domain hash scheme (FDH). Security of the RSA signature with FDH can be proven by letting the output size of the underlying one-way hash function be the same as the size of the composite. However, secure construction of such a length of one-way hash function has not been studied in detail, and consequently, FDH is not frequently utilized in real systems.

PSS is an alternative method that effectively adds randomness to RSA signature schemes. For the probabilistic signature scheme, there could be two different signatures of the same message. It is also a technique very similar to OAEP for public key encryption schemes in a way that PSS is also a pre-process performed on a message before signing. RSA signature scheme with PSS is provably secure, and is commonly used for the implementation of the RSA cryptosystem in practical communication systems, for example PKCS#1ver2.1.

As for digital signature algorithm (DSA), there are no general attacks that can achieve existential forgery (modification had been done) even if the adversary is able to ask the signer to sign any message at his will. While its security has not yet been proven, it has stood the test of time and is now a widely accepted technique.

References

[1]   Shannon, C. E., "Communication Theory of Secrecy Systems," *Bell System Technical Journal,* Vol. 28, 1949, pp. 656–715.

[2]   Shannon, C. E., "A Mathematical Theory of Communication," *Bell System Technical Journal,* Vol. 27, 1948, pp. 379–423 and 623–656.

[3]   Diffie, W., and M.E. Hellmans, "New Directions in Cryptography," *IEEE Trans. on Information Theory,* Vol. 22, 1976, pp. 644–654.

[4]   Rivest, R., A. Shamir, and L. Adleman, "A Method for Obtaining Digital Signature and Public-Key Cryptosystems," *Communication of the ACM,* Vol. 21, No.2, 1978, pp. 120–126.

[5]   Bellare, M., and P. Rogaway, "Optimal Asymmetric Encryption," *Advances in Cryptology—EUROCRYPT'94, Lecture Notes in Computer Science 950,* Springer-Verlag, 1994, pp. 92–111.

[6]   ElGamal, T., "A Public Key Cryptosystem and a Signature Scheme Based on Discrete Logarithms," *IEEE Trans. on Information Theory,* Vol. IT-31, No. 4, 1985, pp. 469–472.

[7]   Rackoff, C., and D.R. Simon, "Non-Interactive Zero-Knowledge Proof of Knowledge and Chosen Ciphertext Attack," *Advances in Cryptology—CRYPTO'91, Lecture Notes in Computer Science 576,* Springer-Verlag, 1992, pp. 433–444.

[8]   Bellare, M. et al., "Relations Among Notions of Security for Public-Key Encryption Schemes," *Advances in Cryptology—CRYPTO'98, Lecture Notes in Computer Science 1462,* Springer-Verlag, 1998, pp. 26–45.

[9]   Bleichenbacher, D., "Chosen Ciphertext Attacks Against Protocols Based on the RSA Encryption Standard PKCS #1," *Advances in Cryptology—CRYPTO'98, Lecture Notes in Computer Science 1462,* Springer-Verlag, 1998, pp. 1–12.

[10] Cramer, R., and V. Shoup, "A Practical Public Key Cryptosystem Provably Secure Against Adaptive Chosen Ciphertext Attack," *Advances in Cryptology— CRYPTO'98, Lecture Notes in Computer Science 1462*, Springer-Verlag, 1998, pp. 13–25.

[11] Fujisaki, E., and T. Okamoto, "Secure Integration of Asymmetric and Symmetric Encryption Schemes," *Advances in Cryptology—CRYPTO'99, Lecture Notes in Computer Science 1666*, Springer-Verlag, 1999, pp. 537–554.

[12] Goldwasser, S., S. Micali, and R. Rivest, "A Digital Signature Scheme Secure Against Chosen-Message Attacks," *SIAM J. on Computing*, Vol. 17, 1988, pp. 281–308.

[13] Coron, J.S., D. Naccache, and J. Stern, "On the Security of RSA Padding," *Advances in Cryptology—CRYPTO'99, Lecture Notes in Computer Science 1666*, Springer-Verlag, 1999, pp. 1–18.

[14] Bellare, M., and P. Rogaway, "The Exact Security of Digital Signatures—How to Sign with RSA and Rabin," *Advances in Cryptology—EUROCRYPT'96, Lecture Notes in Computer Science 1070*, Springer-Verlag, 1996, pp. 399–416.

3

Security Features in Wireless Environment

3.1 Introduction

Security is a critical issue in mobile radio applications both for the users and providers of such systems. Although the same may be said of all communications systems, mobile applications have special requirements and vulnerabilities, and are therefore of special concern. Wireless networks share many common characteristics with traditional wire-line networks such as public switch telephone/data networks, and therefore, many security issues with the wire-line networks also apply to the wireless environment. Wireless networks, while providing many benefits over their wired counterparts, including the elimination of cabling costs and increased user mobility, present some serious security concerns. Unlike wired networks, where the physical transmission medium can be secured, wireless networks use the air as a transmission medium. This allows easy access to transmitted data by potential eavesdroppers. The mobility of wireless networks also introduces problems. The mobility of users, the transmission of signals through the open-air and the low power consumption of the mobile user bring to a wireless network a large number of features distinctively different from those seen in a wire-line network. Issues of security and privacy become more prominent with wireless networks.

3.2 Mobile Network Environment

A simple mobile environment is shown in Figure 3.1.

Generally the following components are found in the mobile network environment:

- Mobile station (MS): A mobile terminal or mobile station is the equipment used by a client to obtain service from the mobile network. If he is within the coverage range of his mobile service provider, he can connect to the mobile network through the cell antenna using his mobile terminal.

- Cell antenna: Cell antennas provide network service facility within its coverage range to the mobile stations.

- Base station controller (BSC): Base station controller is commonly known as base station. It controls a cluster of cell antennas, and is responsible for setting up calls with the mobile station. Also, when the mobile station moves from

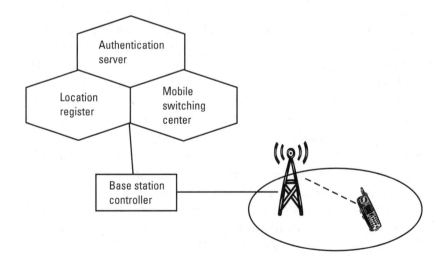

Figure 3.1 Mobile network.

one cell antenna coverage area to another, the base station manages the handoff process to maintain the continuity of the call. The hand off process is transparent to the mobile user.

- Mobile switching center (MSC): An MSC connects all base stations and passes messages and communication signals to and from mobile stations operating on the network.

- Location register: The location register contains information related to the location and the subscription of the users in its domain.

- Authentication server: An authentication server is present in every domain, and stores all confidential information, such as keys, and is assumed to be physically protected.

BSC, MSC, location register, authentication server – all these components of a mobile network can be integrally considered as domain servers. If a mobile station registers with a server, then the server in question becomes its home server and the network belonging to the home server is the home network. The domain administered by the home server is called the home domain. The mobile station can move from one place to another within the home domain or move outside its home domain to a "visiting" domain. The serving network is the one that is currently providing service in the area where the user has roamed. Typically, a serving network has to query the user's home network for information about the user for security and authentication purpose.

Mobile entities use air interface to communicate with the server in order to obtain the services as shown in Figure 3.2. The air interface is vulnerable to both active and passive attacks. An active attacker can subvert the communications between the communicating honest entities by injecting, deleting, altering, or replaying messages. A passive attacker can eavesdrop on the communication link to acquire knowledge of the communications.

The wireless medium is intrinsically a broadcast-based medium. An eavesdropper is able to tap into the wireless communications channels by positioning himself anywhere within the area of the cell.

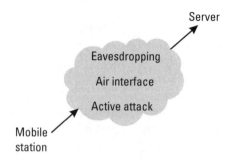

Figure 3.2 Mobile environment.

Since all transmitted data travel directly between a mobile host and the base station, it is possible to copy all the data of a particular message transmitted through the air.

It is also harder to control visiting hosts overloading the network with excessive transmissions, resulting in a sudden decrease in network performance. This may lead to denial of service to other mobile hosts because of the congested network.

There is a security threat during channel setup. When a mobile host "pops-up" in a cell, the base station (or any other network entity carrying out network management tasks and has jurisdiction over that cell) needs to update information on the network in order to allow messages to be routed to that mobile host correctly. This means that information on the physical location of the mobile host are available to entities that are able to see this routing information, an undesirable situation if that mobile user prefers to keep the location private. An impostor may also able to monitor that mobile user and begin connecting to the network using that mobile user's identity after a disconnection. The impostor will then have access to all the resources that are available to the real user. The real user may even be denied connection later because the base station might think that it is trying to reconnect again for the second time.

3.3 General Security Threats of a Network

A network environment is, in general, susceptible to a number of security threats. These include the following:

Masquerading   By masquerading, an entity can obtain unauthorized privileges. In a network system, a masquerading user or host may deceive the receiver about its real identity.

Unauthorized Use of Resources   The unauthorized user can access the network system and utilize the resources for its own purpose.

Unauthorized Disclosure and Flow of Information   This threat involves unauthorized disclosure and illegal flow of information stored, processed, or transferred in a network system both internal and external to the user organizations.

Unauthorized Alteration of Resources and Information   Unauthorized alteration of information may occur both within a system and over the network. This attack may be used in combination with other attacks, such as a replay, whereby a message or part of a message is repeated intentionally to produce an unauthorized effect. This threat may also involve unauthorized introduction, including removal of resources from or into a distribution system.

Repudiation of Actions   This is a threat against accountability in organizations. For instance, a repudiation attack can occur whereby the sender (or the receiver) of a message denies having send (or received) the information.

Unauthorized Denial of Service   Here the attacker acts to deny resources or services to entities authorized to use them. In the case of a network, the attack may involve blocking access to the network by continuous deletion or generation of messages, causing the target to be either depleted or saturated with meaningless messages.

3.4   Limitations of Mobile Environment

Mobile devices are designed to be portable (i.e., light and small). Until a more suitable alternative is found, mobile devices will more than likely continue to be battery powered in the foreseeable future. The power consumption of mobile devices directly affects their usage time. In order to conserve energy, both processing speeds and processor cycles need to be reduced. Because data transmission also consumes energy, it, too, should be reduced. The former imposes limits on the computational complexity of the encryption algorithms and the number of messages involved in security protocols. In addition, integration of security features into mobile devices must take into account applicable restrictions such as small packet size, low bandwidth, high transmission costs, limited processing and storage resources, and real time constraints.

3.5   Mobility and Security

The mobile environment aggravates some of the above security concerns and threats because the security of wireless communication can be compromised much more easily than that of wired communication.

The situation gets further complicated if the users are allowed to cross security domains. (For details on domain boundary crossing, see Section 3.7.4.)

Being reachable at any location at any time creates greater concern about privacy issues among the potential users. For instance, there may be a need for developing profiles that specify who, when, and from is authorized to get a service.

Mobile users will use resources at various locations, provided by various service providers. It is important to understand the trust issues involved when mobile clients are allowed to use resources of different servers at different locations.

Integrity and confidentiality of information stored on the mobile appliance is another important concern. Needless to say, user anonymity [1] is important in a mobile environment. Different degrees of anonymity can be provided, such as hiding user identity from eavesdroppers or from certain administrative authorities.

3.6   Attacks in Mobile Environment

Many ingenious attacks have been developed for compromising security protocols. The results of these attacks, if successful, can range from a mild inconvenience to a severe breach of security. Even when the attacks are unsuccessful they can consume the processing resources of the attacked party and thus reduce the resources available to legitimate communication.

A general problem with wireless communications is that attacks broadcast over the network are difficult to prevent. In a wired network, the attacker must physically "tap" into a wire in the network. Standard security measures can be taken to reduce the access to network wires, such as restricted building access or locked communication closet, and upon detecting and locating a tap, it can be easily removed.

This same property does not exist in a wireless network. Any party that possesses the proper equipment, whether a legitimate member of the network or not, can receive and send messages in the network. When the attackers are discovered it is difficult to purge them from the network because they can roam freely throughout the wireless region while attacking at will.

The attacks described in the following sections are particularly troublesome in wireless communications because they are easy to execute yet impose significant overhead on user or the wireless network. Remedies for each of these attacks are also discussed.

3.6.1 Nuisance Attack

The primary concern of security protocol is to successfully defend against all forms of attacks. However, even when an attack is thwarted, it has typically required the entity or the server to expend processing and communication resources to discover the attack. These attacks are referred to as nuisance attacks, because while they cannot compromise security, they can disrupt the activities of legitimate users. This disruption can cause significant problems in a wireless network since the mobile terminal typically consists of minimal processing resources. Therefore, the resulting attacks can severely affect the mobile principal's ability to conduct legitimate communications. Of greater concern are the communications bandwidth and cost. A nuisance attack introduced into a wireless network can result in several unnecessary wireless responses being expended before the attacking message is discovered to be fraudulent. Protocol design should try to minimize the possibility or at least the impact of the nuisance attack.

3.6.2   Impersonation Attack

By impersonating a legitimate user, the attacker will try to eliminate one of the communicating parties from the intended communication, making the other communicating party believe that he is legitimate. The attacker can impersonate the mobile entity and with a high computational ability, it can also impersonate the server. In order to avoid impersonation attack, the protocol design should consider the mutual authentication of the legitimate entity and the server.

3.6.3   Interception Attack

In wireless communication, communicating messages are transmitted by air, allowing someone to easily tap into the communication without being detected, and access all communicating messages. In a wireless environment, it is not possible to eliminate interception attack, but by encrypting all the messages in communication, it is possible to prevent the attacker from gaining any valuable information.

3.6.4   Replay Attack

Using this method, the attacker intercepts and stores all communications between the communicating parties. At a later time, the attacker impersonates one of the communicating parties by replaying the stored messages. By incorporating the session variant parameter in authentication messages, it is possible to resist replay attacks.

3.6.5   Parallel Session Attack

Using this attack method, the attacker begins communication with one of the communicating parties and uses it as an oracle to compute the session key. This method is most successful when the flow of messages between the communicating parties is of the same structure. These attacks can be effectively prevented by maintaining asymmetry in the back and forth messages and by including direction dependent parameters in the message flows.

Bird et al. [2] identified another form of oracle attack, one in which an adversary starts two separate authentications sessions with the server and user. When interacting with the server, it becomes the user, and vice versa. It tries to take the advantage of the messages from the authentication session with the server to impersonate the server in authentication session with the user. This kind of attack can be effectively prevented if the encrypted messages used in each run of the protocol are different from, or logically linked, with one another.

3.7 Security Issues in Mobile Environment

In the following sections, four major areas of mobile systems security are discussed, namely *authentication, anonymity, device vulnerability,* and *domain crossing.*

3.7.1 Authentication

The primary objective of an authentication scheme is to prevent unauthorized users from gaining access to a protected system [3]. As with current distributed systems, authentication is a necessary procedure for verifying both an entity's identity and authority. The level of trust for a particular entity depends on the outcome of this authentication process. Ideally, user authentication should be carried out transparently, without disruption to whatever the user's current task. Authentication protects the service provider from unauthorized intrusion. By mutual authentication [4] mobile station also authenticates the server. There are two reasons why this could be of importance. First, it prevents a malicious station from pretending to be a base station. Then it permits the MS to choose the services of a particular base station in the presence of colocated networks.

In practice, most authentication protocols require the home *authentication authority* (or authentication server) to be contacted during or before the execution of the protocol. Consider the overhead that will be incurred when this has to be done for many mobile users entering the foreign domain. Furthermore, the "transparency" requirement for authentication protocols would be difficult to meet. The completion time for each protocol also depends on the quality of the link between the visited domain and the mobile user's home authentication server. This also means that the home authentication server must be available at all times. These last two factors, the link quality between the visited domain and the user's home authentication server and the availability of the authentication server itself, are unpredictable and therefore cannot be guaranteed.

While the use of certificates may relax the requirement of contacting the user's home authentication server, it also contains some undesirable properties. For one, it is irrational to assume that the certifying authority signing the certificate is globally and unconditionally trusted by every entity. Also, a mobile user may travel from one domain to other domains and it may not always be preplanned. In these instances, it is not possible for the user to obtain certificates issued by his home domain for the all other domains he might visit and the visiting domain authority may not accept the certificate.

Another problem is that certificates do not reflect the *current status* of its owner/carrier (e.g., the current balance of a bank account or a record of his or her behavior in previously visited domains). It is difficult to embed some information about the current status of the user into the certificate by the server and at the same time be sure that the user cannot alter that information or present only certificates that provide the most positive credentials. Revocation of certificates will also become a more difficult problem, and one concerned with scalability, in that mobile users move frequently, and their locations could be anywhere in the world.

Engineering good authentication protocols for mobile systems carry an extra burden of anonymity requirements. It is imperative that authentication protocols release as little information as possible relating to the principals involved in the protocol execution.

3.7.2 Anonymity

Anonymity is the state of being not identifiable within a set of principles [5]. Information about a particular person or organization is private and should only be known to its owner and to whomever he grants access rights. Privacy should be preserved in any kind of information system, be it fixed or mobile. The type of information that a user may want to keep private could include his real user identity when on-line, his activities, his current location and his movement patterns. Preserving anonymity [1] is of greater concern in mobile systems for several reasons. Mobile systems yield more easily to eavesdropping and tapping, compared to fixed networks, making it easier to tap into communication channels and obtain user information. As users move around, a new kind of information immediately becomes valuable, such as detailed information about the movement and location of the user. This may also provide clues to any user interaction at a given point in time. Users will also move in and out of foreign domains without the prior knowledge of the user, and therefore may not be completely trustworthy. Moving across foreign domains thus results in increased risk to user information. Current network implementers of mobile communication systems store a lot of user related information on network databases, especially for mobile telecommunication networks. This is done to assist in user mobility support as well as billing and authentication. This makes the user information more widespread and highly available. It is also uncertain whether the environment where this data is stored is safe and trustworthy. The following issues should be considered to solve the anonymity problem:

- Preventing other parties from making any association of the user with messages that he or she sent or received;

- Preventing any association of the user with communication sessions in which he or she may participate;

- Preserving the privacy of location and movement information of users;

- Preventing the disclosure of the relationship between a user and his or her home domain;

- Preventing any association of the user with the foreign domains he or she may have visited;

- Disallowing the exposure of a user's activities, by hiding his or her relationship with the visited domains.

Users can be denied service by various mechanisms, usually by either "cutting off" the communication channel between the client and the server or by flooding the network to the extent that no more bandwidth is available for use, rendering the network effectively nonoperational. With *unselective* denial of service, whole services or large parts of a network are disabled (e.g., using explosives), and these are usually detectable. *Selective* denial is less evident and its victims are usually well defined (e.g., a particular client on the network). Anonymity is an obvious solution to the latter problem.

A common solution that has been adopted, providing a certain degree of anonymity in current systems, is by means of an *alias,* or a temporary identity. Aliases or nicknames allow a user to be referenced without revealing his real identity. Another way to provide user anonymity is to encrypt the real identity [6].

3.7.3 Device Vulnerability

Mobile devices are designed to be small and lightweight, making them highly portable. These features of mobile devices make them potentially vulnerable to being misplaced or lost, and worse, to theft. Even though losing the physical device itself is an unsatisfactory enough outcome, a more detrimental consequence is the owner's deprivation of the information or data that is (or was) contained in his or her device. Hardware can be repurchased, but information, especially the kind that is updated frequently, cannot be refabricated that easily. Worse still, some of the data may contain a secret not even known to the owner.

Mobile devices may also be used as a control device. Examples include active badges for controlling access to workstations and building entrances and even devices used when purchasing goods or withdrawing money (e-cash) from an ATM. Without these devices, the users will be denied access to most of these facilities and services. Furthermore, if procedures for obtaining a replacement device of this type take time to process, the device owner's industrial and social progress will be severely affected.

If the device is stolen, thieves who can disarm the safety features on the device can then access the private information contained within. The thief may also get unauthorized access to services that are available via that device, prior to the theft being discovered and privileges are revoked.

3.7.4 Domain Boundary Crossing

A security domain means a set of network entities on which a single security policy is employed by a single administrative authority.

Security domain boundaries are crossed when a mobile user leaves one security domain and enters another.

Upon entering a new domain, the trustworthiness of the new domain environment has to be ascertained by the mobile user, and vice versa. This is usually carried out using mutual authentication protocols where two entities mutually authenticate each other during one protocol execution. It is important at this stage to determine the *trustworthiness* of the domain and user. The level of trust established will form the basis on which security related activities and decisions are made.

Another important motivation for domains to screen its visiting hosts is to uphold its image as a *safe* domain. Much like geographic domains (e.g., cities or suburbs), a hostile environment will tend to be avoided and its resident occupants would want to migrate to a safer haven. The consequences would lay the economical soundness of that domain, among other activities, in jeopardy.

References

[1] Samfat, D., R. Molve, and N. Asokan, "Untreacibility in Mobile Networks," *Proc. of ACM Int. Conf. on Mobile Computing and Networking*, Berkeley, CA, November 1995.

[2] Bird, R., et al., "Systematic Design of a Family of Attack-Resistant Authentication Protocols," *IEEE Journal on Selected Areas in Communications*, Vol. 11, No. 5, 1993, pp. 679–693.

[3] Morris, R., and K. Thompson, "Password Security: A Case History," *Communications of the ACM*, Vol. 22, No. 11, 1997, pp. 594–597.

[4] Joos, R. R., and A. R. Tripathi, *Mutual Authentication in Wireless Networks*, Technical Report, Computer Science Department, University of Minnesota, 1997.

[5] Pitzmann, A., and M. Köhntopp, "Anonymity, Unobservability, and Pseudonymity—A Proposal for Terminology," *Designing Privacy Enhancing Technologies, LNCS 2009*, Springer-Verlag, 2001, pp. 1–9.

[6] Park, C. S., "Authentication Protocol Providing User Anonymity and Untreacibility in Wireless Mobile Communications Systems," http://www.misecurity.com/ko/forum/ forum_06.pdf.

4

Standard Protocols

This chapter provides an overview of wireless networking protocols and their standards. IEEE standardized IEEE 802.11 is the standard for wireless LAN system and Bluetooth is the standard for wireless ad hoc networks.

4.1 IEEE 802.11

4.1.1 Brief History

Motorola developed one of the first commercial wireless local area network (WLAN) systems with its Altair product. Altair was designed mostly to proprietary RF (radio frequency) technologies, provided low data rates and was prone to radio interference. In 1990, the Institute of Electrical and Electronics Engineers (IEEE) initiated the 802.11 project with a scope to develop a medium access control (MAC) and physical layer (PHY) specification for wireless connectivity for fixed, portable, and moving stations within an area. Seven years later, on June 27, 1997, IEEE first approved the IEEE 802.11 international interoperability standard for WLAN. IEEE 802.11 standard uses 2.4 GHz ISM (industrial, scientific, and medical) radio band and provides a mandatory 1 Mbps and an optional 2 Mbps data transfer rate. In 1999, IEEE ratified the 802.11a and the 802.11b wireless networking communication standards. IEEE 802.11b standard operates in the 2.4 ~2.5 GHz ISM band and permits transmission speed up to 11 Mbps. The 802.11a standard is a high-speed interface definition that can produce data at up to 54 Mbps and operates in the 5-GHz frequency spectrum.

4.1.2   IEEE802.11 Architecture

Mobile nodes connect to the fixed network through the fixed access point (AP) on a wired network, allowing the establishment of a peer-to-peer connection. The standard defines two types of wireless network topologies: one is infrastructure mode, which IEEE standard defines as basic service set (BSS) and the other one is ad hoc mode defined as independent basic service set (IBSS).

4.1.2.1   Infrastructure Network

This network architecture extends the range of wired LAN to wireless cells. A stationary device, is a part of the wired network infrastructure, and provides network connectivity to the portable or mobile devices. The AP is actually the bridging point between a wired network and a wireless network. A cell is the area covered by an AP. If PCs, or laptops with a wireless network interface card (NIC), or other mobile devices with wireless network connectivity are within the range of an AP, then they can get the network connection. Figure 4.1 shows the BSS topology of an 802.11 WLAN.

The collection of all BSS of an infrastructure network is called an extended service set (ESS). Figure 4.2 depicts the ESS infrastructure. The BSS network is limited in its range. If the coverage areas of different BSS overlap each other then it is possible to move from one cell to another without losing network connection, and thus a broad network coverage can be achieved. Thus, this can be used to replace the wired connectivity within a building or campus areas. This structure is also used to extend LAN infrastructure.

The 802.11 standard recognizes the following mobility types:

- *No-transition.* This type of mobility refers to stations that do not move and those that are moving within a local BSS.

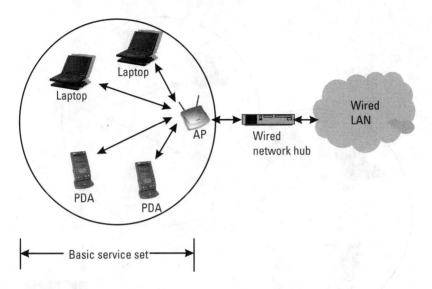

Figure 4.1 BSS topology.

- *BSS-transition.* This type of mobility refers to stations that move from one BSS in one ESS to another BSS within the same ESS.

- *ESS-transition.* This type of mobility refers to stations that move from a BSS in one ESS to a BSS in a different ESS.

Within the ESS, the 802.11 standard accommodates the following physical configuration of BSSs:

- *BSSs partially overlapped.* This type of configuration provides contiguous coverage within a defined area, which is best if the application cannot tolerate a disruption of network service.

- *BSSs physically disjointed.* For this case, the configuration does not provide contiguous coverage. The 802.11 standard does not specify a limit to the distance between BSSs.

- *BSSs are physically collocated.* In this case, it may be necessary to provide a redundant or higher-performing network.

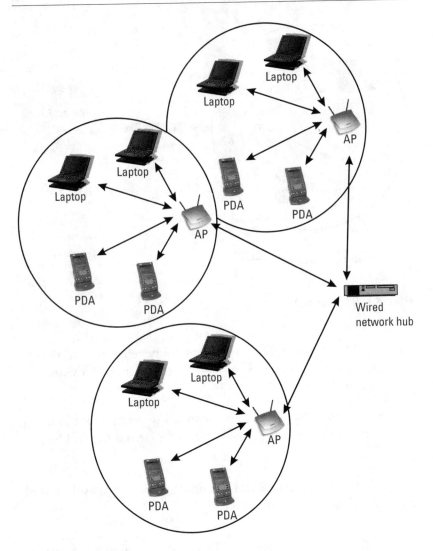

Figure 4.2 ESS topology.

4.1.2.2 Ad Hoc Network

This architecture is completely wireless in that there is no AP to inter-link with the wired network. In ad hoc mode each client communicates with the other clients within the network. Ad hoc mode is designed such that only the client stations within the transmission range (within the same cell) of each other can communicate. The ad hoc topology is shown in Figure 4.3. If a client in an ad hoc network

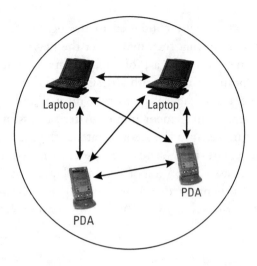

Figure 4.3 IBSS topology.

wishes to communicate outside the cell, a member of the cell must operate as a gateway and perform routing.

Prior to communicating data, wireless clients and APs have to establish a relationship called association. Only after an association is established, can two clients communicate. In infrastructure mode, the clients associate with the AP. The association process has three states:

1. Unauthenticated and unassociated;
2. Authenticated and unassociated;
3. Authenticated and associated.

In order to change the association states, the communicating parties exchange messages, which are called management frames. The process is as follows.

All APs transmit a beacon management frame at a fixed interval. If the client station is within the range of an AP (i.e., within a BSS), it can pay attention to the beacon. If it is within the range of multiple APs overlapping coverage area, then it listens to all the beacon messages transmitted by different APs. The client can then select the BSS to join in a vendor independent manner. For instance, on the Apple Macintosh, all of the network names (or the service set identifiers

(SSID)) which are usually contained in the beacon frame are presented to the users so that they may select the network they wish to join. A client may also send a probe request management frame to find an access point affiliated with a desired SSID. After identifying an AP, the client and the AP perform a mutual authentication by exchanging several management frames. After successful authentication, the client moves into the second state, authenticated and unassociated. The client then sends an association request and AP responds with an association response frame. The client is now in the third stage, the authenticated and associated. The client now becomes a peer on the wireless network and can communicate with the network.

4.1.3   IEEE 802.11 Layers

The IEEE standard places specifications on the parameters of data link layer and physical layer of the OSI model. Figure 4.4 shows the 802.11 and the OSI model. The standard defines a general media access control (MAC) layer and three different PHY-layers. The MAC-layer is the same for all the three PHY-layers.

4.1.3.1   IEEE 802.11 MAC Layer

The goal of the MAC layer is to provide access control functions for shared-medium PHY-layers in support of the logical link control (LLC) layer. The MAC layer performs the addressing and recognition of frames in support of the LLC. The 802.11 standard uses carrier sense multiple access with collision avoidance (CSMA/CA), similar to that in Ethernet. Standard Ethernet uses carrier sense multiple access with collision detection (CSMA/CD). In the CSMA/CA protocol, when a node wants to send some data packet, it first listens to ensure no other node is transmitting. If the channel is clear, it then transmits the packet. Otherwise, it chooses a random "backoff factor" that determines the amount of time the node must wait until it is allowed to transmit its packet. During periods in which the channel is clear, the transmitting node decrements its backoff counter (when the channel is busy, it does not decrement its backoff counter). When the backoff counter reaches zero, the node transmits the packet. Since the probability that two nodes will choose the same backoff factor is

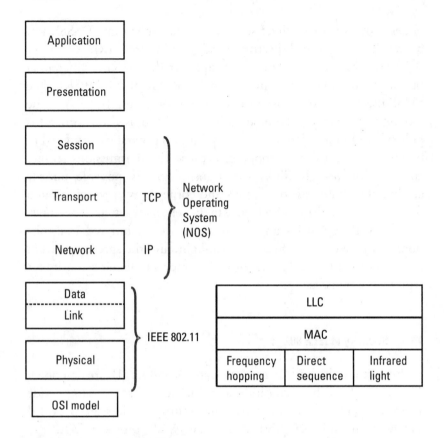

Figure 4.4 OSI model and IEE802.11.

small, collisions between packets are minimal. The loss of speed due to the random waiting time is compensated by the lesser retransmissions.

Collision detection, as is employed in Ethernet, cannot be used for the radio frequency transmissions of IEEE 802.11. The reason for this is that when a node is transmitting, it cannot hear any other node in the system which may be transmitting, since its own signal will drown out any others arriving at the node.

4.1.3.2   IEEE 802.11 PHY-Layer

The PHY- layer specifies the modulation scheme used and signaling characteristics for the transmission through the radio frequencies. PHY-layer, which actually handles the transmission of data between nodes, can use either direct sequence spread spectrum (DSSS), frequency hopping spread spectrum (FSSS), or infrared (IR).

The chosen modulation technique for the DSSS is differential bi and quadrature phase shift keying (DBPSK and DQPSK). The FHSS uses 2-4 level Gaussian frequency shift keying (GFSK) as the modulation scheme. The modulation technique used for infrared is pulse position modulation (PPM). Infrared is generally considered to be more secure to eavesdropping, because IR transmissions require absolute line-of-sight links (no transmission is possible outside any simply connected space or around corners), as opposed to radio frequency transmissions, which can penetrate walls and be intercepted by third parties unknowingly. However, infrared transmissions can be adversely affected by sunlight, and the spread-spectrum protocol of IEEE 802.11 does provide some rudimentary security for typical data transfers.

4.1.4   Security of IEEE 802.11

In this section, the built-in security features of 802.11b are described. The IEEE 802.11b specification identifies several services to provide a secure operating environment. The security services are provided largely by the Wired Equivalent Privacy (WEP) protocol. WEP was part of the original IEEE 802.11 wireless standard. WEP protocol is used only to protect link level data during wireless transmission between clients and the access points. WEP provides security for the wireless portion of the connection, but does not provide end-to-end security. Security of a typical network is shown in Figure 4.5.

The basic security services provided by IEEE 802.11b are:

Authentication. The primary goal of WEP is to provide access to the legitimate clients. The network has to verify the identity of the communicating client station. That is done to provide access control to the network through denying access to the client stations that cannot authenticate properly.

Data confidentiality. The second goal of WEP is to provide data confidentiality. The goal is to prevent data compromise by eavesdropping (passive attack). Data is protected by enciphering

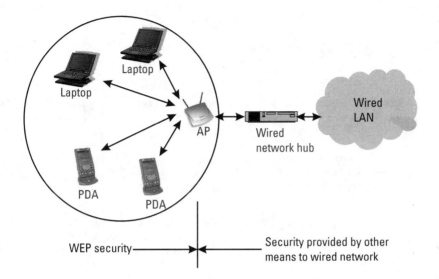

Figure 4.5 Security of a typical network.

them and allowing decryption only by clients who have the correct WEP key.

Data integrity. Another goal is to ensure that data is not modified in transit between the wireless clients and the access point in an active attack.

4.1.4.1 Authentication

The IEEE 802.11b specification defines two means of an authentication process. One is cryptography based, and the other is noncryptographic. The noncryptographic approach is an identity-based verification. There are also two different approaches in identity-based verification—open system authentication and closed system authentication. In both cases, the wireless client requests access simply with the service set identifier (SSID) of the wireless network. The classification of authentication technique is shown in Figure 4.6.

Open System Authentication

This is the default authentication protocol for IEEE 802.11. As the name implies, open system authentication provides access to anyone who wants to get connected to the network. To obtain access, a client

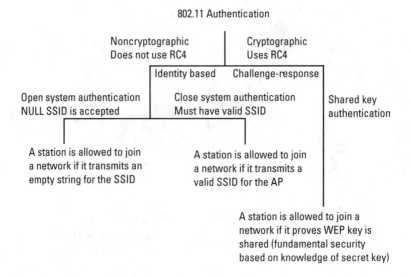

Figure 4.6　Classification of IEEE 802.11b authentications.

has to send only a null string for the SSID. Hence, it provides a null authentication process.

Closed System Authentication

In a closed network, only those clients with the knowledge of the network name, or SSID, can join. To get access, the wireless client must respond with the actual SSID of the wireless network. Thus, a client is allowed access if it responds with the correct 0- to 32-byte string identifying the access point of the wireless network.

Both open and closed system authentication schemes simply provide identification, as practically, there is no true authentication. Both open and closed authentication schemes are highly vulnerable to attacks against even the most novice adversaries.

Shared Key Authentication

Shared key authentication is a cryptographic authentication which uses a simple challenge-response scheme based on whether a client has the knowledge of a shared secret, such as a key. The initiator, the wireless client wishing to authenticate, sends an authentication request management frame indicating that it wishes to use shared key authentication.

After receiving the authentication request, the access point (AP)/BSS, the responder, responds by sending an authentication management frame (Figure 4.8) containing 128 octets of challenge text to the initiator. The challenge text is generated by using the WEP pseudorandom number generator (PRNG) with the shared secret key and a random initializing vector (IV). Once the initiator receives the management frame, it copies the content of the challenge text into a new management frame body. This new management frame body is then encrypted with WEP using the shared secret along with a new IV selected by the initiator. The encrypted management frame is then sent to the responder. The responder decrypts the received frame and verifies that the 32-bit CRC integrity check value (ICV) is valid, and that the challenge text matches the one sent in the challenge message. If they do, then authentication is successful. If the authentication is successful, then the AP and the client switch their role as initiator and responder respectively and repeat the process to ensure mutual authentication. The entire process is shown in Figure 4.7 and the format of an authentication management frame is shown in Figure 4.8. The format shown is used for all authentication messages.

The value of the status code field is set zero when successful, and to an error value if unsuccessful. The element identifier confirms that

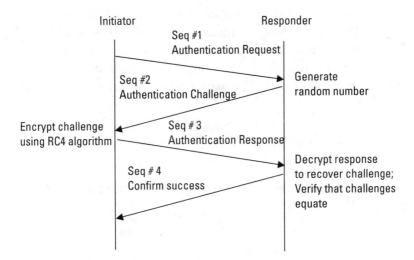

Figure 4.7 Shared key authentication message flow.

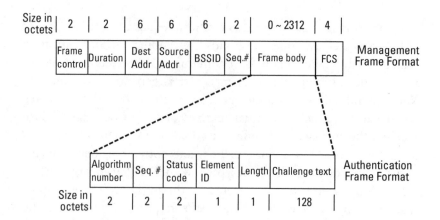

Figure 4.8 Authentication management frame.

the challenge text is included. The length field identifies the length of the challenge text and is fixed at 128. The challenge text includes the random challenge string. Table 4.1 shows the possible values and when the challenge text is included based on the message sequence number.

4.1.4.2 Data Confidentiality

In a wireless environment, when transmissions are broadcast over the radio link, interception and masquerading becomes trivial to anyone with a radio. IEEE 802.11 standard introduced the Wired Equivalent Privacy (WEP) protocol to protect authorized uses of wireless LAN from eavesdropping and bring the security level of the wireless systems closer to that of the wired counterpart. A secondary function of WEP is to prevent unauthorized access to a wireless network; this

Table 4.1
Message Format Based on Sequence Number

Sequence Number	Status Code	Challenge Text	WEP Used
1	Reserved	Not present	No
2	Status	Present	No
3	Reserved	Present	Yes
4	Status	Not present	No

function is not an explicit goal in the IEEE 802.11 standard, but it is frequently considered to be a feature of WEP.

WEP is intended to provide functionality for the wireless LAN; equivalent to that provided by the physical security attributes inherent to a wired medium. WEP uses RC4 symmetric key stream cipher algorithm to generate encrypted data. Through the use of WEP technique, data can be protected from disclosure during transmission over the wireless link. WEP is applied to all data above the 802.11 WLAN layers to protect traffic such as Transmission Control Protocol/ Internet Protocol (TCP/IP), Internet Packet Exchange (IPX), and Hypertext Transfer Protocol (HTTP).

Protocol Description

WEP is a symmetric key algorithm in which the same key is used for encipherment and decipherment. The encrypted packet is generated with a bitwise exclusive OR (XOR) of the original plaintext with a pseudorandom key sequence of equal length. WEP supports cryptographic keys sizes from 40 to 104 bits. However in practice most WLAN deployments rely on 40-bit key. Figure 4.9 shows the enciphering process of the WEP algorithm. A secret key has been

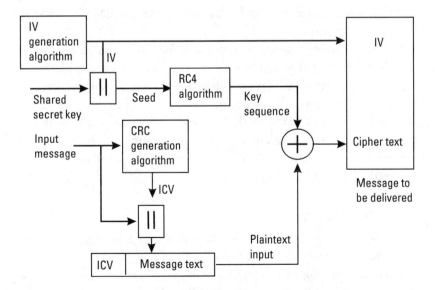

Figure 4.9 WEP enciphering process.

distributed to the wireless stations by an external key management service. WEP relies on this secret key shared between the communicating parties to protect the body of a transmitted frame of data. The enciphering process is as follows.

First, an integrity algorithm (CRC approach) operates on the input message (payload) to produce an integrity check value (ICV). ICV is then concatenated with the message text to produce the plaintext input to be encrypted. IV generation algorithm generates a 24-bit initialization vector (IV). The secret key is concatenated with IV and the resulting seed is input to RC4 algorithm. The RC4 algorithm outputs a key sequence of equal in length to the plaintext input. Encipherment is then performed by bitwise XORing the key sequence with the plaintext input. The output (message to be delivered) of the whole process consists of the IV and cipher text.

RC4 algorithm is the critical component of this process, since it transforms a relatively short secret key into an arbitrarily long key sequence. This greatly simplifies the key distribution as only the secret key needs to be communicated between the stations. The secret key remains constant while the IV changes periodically. Each new IV results in a new seed and key sequence, thus there is one-to-one correspondence between IV and the key sequence. The IV is transmitted in clear, since its value must be known by the recipient in order to perform decryption.

WEP deciphering starts with the arrival of the message. Figure 4.10 depicts the WEP deciphering process. The IV of the incoming message is concatenated with the shared secret key to generate the key sequence to decipher the incoming message. Produced key sequence is then bitwise XORed with the received ciphertext, resulting in the plaintext output. The plain text output contains the ICV and the output text. The output ICV is used to check the validity of the received message.

4.1.4.3   Data Integrity

IEEE 802.11 also offers a means to provide data integrity for messages transmitted between wireless client and access points. This security service was designed to reject any message that has been modified by an active adversary "in the middle." WEP uses simple cyclic redundancy check (CRC) approach to provide data integrity. As

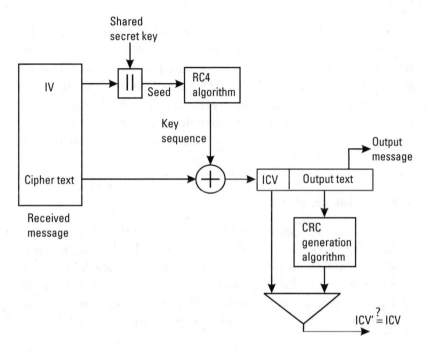

Figure 4.10 WEP decryption and integrity check.

shown in Figure 4.10, a 32-bit ICV is computed on each payload and ciphertext is generated by XORing RC4 key stream with the concatenated ICV and payload. On the receiving end, decryption is performed. Output of the decryption process is the concatenated ICV and text output. The output text is then passed through the CRC generation algorithm and the computed ICV′ is then compared with the deciphered ICV. If the ICVs do not match, then this would indicate an integrity violation and the received message would be discarded. Unfortunately, IEEE 802.11 integrity is vulnerable to certain attacks regardless of key size.

4.1.5 Key Management

Key management for IEEE 802.11 is largely left as an exercise for the vendors. The 802.11 standard, however, provides two methods for using WEP keys. The first provides an array of four keys. A wireless station or an AP can decrypt packets enciphered with any one of the four keys. Transmission, however, is limited to the default key, one of the four manually entered keys. The second method is called a key mapping table. In this method, each unique MAC address can have a separate key. The size of a key mapping table should be at least ten entries according to the IEEE 802.11 specification. The maximum size, however, is likely chip-set dependent. The use of a separate key for each user mitigates the cryptographic attacks, but enforcing a reasonable key period remains a problem as the keys can only be changed manually.

4.1.6   Weaknesses of WEP

WEP has defenses against data integrity and confidentiality. It uses an integrity check value (ICV) to ensure that the data has not been modified during transmission. An initialization vector (IV) is used to augment the shared key and produce a different RC4 key stream for each packet. Unfortunately, both of these critical measures are implemented incorrectly, resulting in serious security vulnerabilities in WEP.

- The use of static WEP keys. Many users in a wireless network potentially sharing the identical key for long periods of time is a well-known security vulnerability. This is in part due to the lack of any key management provisions in the WEP protocol. If a computer such as a laptop were to be lost or stolen, the key could become compromised along with all the other computers sharing that key. Moreover, if every station uses the same key, a large amount of traffic may be rapidly available to an eavesdropper for analytic attacks.

- The IV in WEP is only 24 bits long, which guarantees the reuse of the same IV, and thus, reuse of the same key stream (unless the shared key is changed very frequently, which is rare in current implementations of WEP due to poor key management mechanism). The IV is sent in the clear text portion of a message. As a result, the attacker can actually ascertain that two packets are encrypted with the same key stream. If an attacker flips a bit in the ciphertext the corresponding plain text also gets flipped. By doing statistical analysis on two cipher texts encrypted with the same key stream, the attacker can recover the plain text, including the key stream that was used to encrypt the data.

Let us consider two messages: M_1 and M_2 encrypted with the same key stream K produces ciphertext output C_1 and C_2, respectively:

$$C_1 = M_1 \oplus K$$
$$C_2 = M_2 \oplus K$$

If an eavesdropper gets C_1 and C_2, he or she can perform the following calculation:

$$C_1 \oplus C_2 = (M_1 \oplus K) \oplus (M_2 \oplus K)$$
$$= (M_1 \oplus M_2) \oplus (K \oplus K)$$
$$= (M_1 \oplus M_2)[\text{because } (K \oplus K) = 0]$$

The eavesdropper has now the XOR of two messages, and with the help of simple stochastic algorithm, he or she can access the original messages and, in most cases, the key stream.

- If a key stream for one particular IV is known, then the key stream can be used to inject a new message into the network by XORing a new plain text message with the known key stream. Since 802.11 standard does not require the IV to change with every packet, every device will accept this message as a valid WEP packet.

- The IV is a part of the RC4 encryption key. The fact that an eavesdropper knows 24-bits of every packet key, combined with a weakness in the RC4 key schedule, leads to a successful analytic attack [1], which recovers the key, after intercepting and analyzing only a relatively small amount of traffic. This attack is publicly available as an attack script and open source code.

- WEP provides no cryptographic integrity protection. However, the 802.11 MAC protocol uses a noncryptographic cyclic redundancy check (CRC) to check the integrity of packets, and acknowledge packets with the correct checksum. The CRC computation and XORing the plaintext and keystream are linear computation, so the attacker can make changes in the ciphertext and can generate the CRC part to keep the CRC correct. There is an active attack that permits the attacker to decrypt any packet by systematically modifying the packet and CRC sending it to the AP and noting whether the packet is acknowledged. These kinds of attacks are often subtle, and it is now considered risky to design encryption protocols that do not include cryptographic integrity protection because of the possibility of interactions with other protocol levels that can divulge information about ciphertext.

4.2 Bluetooth

This section provides a detailed overview of ad hoc networks, in particular ad hoc networks based on Bluetooth technology. Ad hoc networks are a relatively new paradigm in wireless communications in which there are no fixed infrastructures, such as base stations or access points. In ad hoc networks, devices maintain arbitrary network configurations formed on the fly, relying on a system of mobile routers connected by wireless links to enable devices to communicate with each other. Devices within the ad hoc network control the network configuration and maintain and share resources.

Ad hoc networks allow devices to access wireless applications, such as address book synchronization and file sharing, within a personal area network (PAN). When combined with other technologies, these networks can be expanded to include network and internet access. Bluetooth devices that typically do not have access to network resources, but that are connected in a Bluetooth network with an 802.11 capable device, can achieve connection within the corporate network as well as reach out to the Internet.

4.2.1 Bluetooth Overview

Ad hoc networks today are based primarily on Bluetooth technology. Bluetooth is an open standard for short range digital radio. It is considered as a low-cost, low-power, and low-profile technology that provides a mechanism for creating a small wireless network on an ad hoc basis. Bluetooth is considered a PAN technology that offers fast and reliable transmission of both voice and data.

Bluetooth can be used to connect almost any device to any other device. An example is the connection between a PDA and a mobile phone. The goal of Bluetooth is to connect disparate devices (PDAs, cell phones, printers, faxes) together wirelessly in a small environment, such as an office or home. According to the leading proponents of the technology (Ericsson, Intel, IBM, and Nokia), Bluetooth is a standard that:

- Eliminates wires and cables between both stationary and mobile devices;

- Facilitates both data and voice communications;

- Offers the possibility of ad hoc networks and deliver synchronicity between personal devices.

Bluetooth transceivers operate in the 2.4 GHz industrial, scientific and medical (ISM) band. Bluetooth transceivers use Gaussian frequency shift keying (GFSK) modulation,and employ a frequency hopping spread spectrum (FHSS) system. The theoretical maximum bandwidth of a Bluetooth network is 1 Mbps. However, in reality the networks cannot support such data rates because of forward error correction (FEC). The second generation of Bluetooth is expected to provide up to 2 Mbps maximum bandwidth.

Bluetooth uses a combination of packet and circuit switching technologies. The advantage of using packet switching in Bluetooth is that it allows devices to route multiple packets of information by the same data path. Since this method does not consume all the resources on a data path, it becomes easier for remote devices to maintain data flow throughout a scatter-net.

4.2.2 Brief History

The original architect for Bluetooth (named after the tenth century Danish king Harald Bluetooth), was Ericsson Mobile Communication. In 1998, IBM, Intel, Nokia and Toshiba formed the Bluetooth Special Interest Group (SIG), which serves as the governing body of the specification. The SIG began as a means to monitor the development of radio technology and the creation of a global and open standard. Today more than 2,000 organizations are part of the Bluetooth SIG comprising leaders in the telecommunications and computing industries that are driving development and promotion of Bluetooth technology. Bluetooth was originally designed, primarily, as a cable replacement protocol for wireless communications. However, SIG members plan to develop a broad range of Bluetooth-enabled consumer devices to enhance wireless connectivity. Among the array of devices that are anticipated are cellular phones, PDAs, notebook computers, modems, cordless phones, pagers, laptop computers, cameras, PC cards, fax machines, and printers. Bluetooth is now standardized within the IEEE 802.15 Personal Area Network (PAN) working group formed in early 1999.

4.2.3 Benefits

Bluetooth offers five primary benefits to users. This ad hoc method of untethered communication makes Bluetooth very attractive results in increased efficiency and reduced costs. The efficiency and cost savings are attractive for both the home user and the enterprise business user.
Benefits of Bluetooth include:

- *Cable replacement:* Bluetooth technology replaces cables for a variety of interconnections. These include peripheral devices such as mouse and keyboard computer connections, USB at 12 Mbps (USB 1.1) up to 480 Mbps (USB 2.0), printers and modems usually at 4 Mbps, and wireless headsets and microphones that interface with PCs or mobile phones.

- *Ease of file sharing:* Bluetooth enables file sharing between Bluetooth-enabled devices. For example, participants in a meeting with Bluetooth-compatible laptops can share files with each other.

- *Wireless synchronization:* Bluetooth provides automatic wireless synchronization with other Bluetooth-enabled devices.

For example, personal information contained in address books and data books can be synchronized between PDAs, laptops, mobile phones, and other devices. The synchronization occurs automatically, without the need of input from the device owner. It automatically occurs whenever the devices come within the range of one another's device transmission, without the device user's knowledge.

- *Automated wireless applications:* Bluetooth supports automatic wireless application functions. Unlike synchronization, which typically occurs locally, automatic wireless applications interface with the LAN and internet. For example, an individual working offline on e-mails might be outside of his or her regular service area, on a flight, for instance. To e-mail the files queued in the inbox of the laptop, the individual, once back in the service area would activate a mobile phone or any other device capable of connecting to a network. The laptop would then automatically initiate a network join by using the phone as a modem and automatically sends the e-mails after the individual logs on.

- *Internet connectivity:* Bluetooth is supported by a variety of devices and applications. Internet connectivity is possible when these devices and technologies join together to use each other's capabilities. For example, a laptop, using a Bluetooth connection, can request a mobile phone to establish a dial-up connection and can then access the internet through this connection.

With all these benefits, Bluetooth is expected to be built into office appliances (e.g., PCs, faxes, printers, laptops), communication appliances (e.g., cell phones, handsets, pagers, headsets) and home appliances (e.g., DVD players, cameras, refrigerators, microwave ovens). Applications for Bluetooth also include vending machines, banking and other electronic payment systems; wireless office and conference rooms; smart home; and in-vehicle communications and parking.

4.2.4 Bluetooth Architecture and Components

Bluetooth network topologies are established on a temporary and ad hoc basis. In this architecture, client stations are grouped into a single geographic area and can be internet worked without access to the wired LAN (infrastructure network). Unlike a WLAN, that comprises both a wireless station and an access point, with Bluetooth there are only wireless stations or clients. A distinguishing feature of Bluetooth networks is the master-slave relationship maintained between the network devices. Up to eight Bluetooth devices may be networked together in a master-slave relationship, called a piconet. In a piconet, as shown in Figure 4.11, one device is designated as the master of the network with up to seven slaves connected directly to that device. The master device controls and sets up the network (including defining the networks hopping scheme). Devices in a Bluetooth piconet operate on the same channel and follow the same frequency hopping sequence. Although only one device may perform as the master for each network, a slave in one network can act as the master for other networks, thus creating a chain of networks. The series of piconets often referred to as scatter-net (Figure 4.12), allows

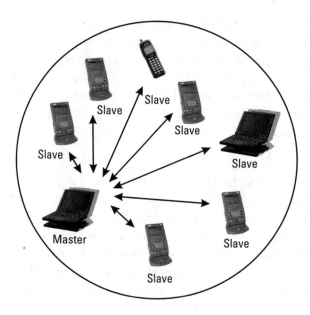

Figure 4.11 Bluetooth piconet.

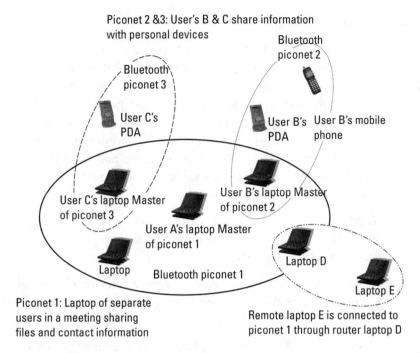

Figure 4.12 Bluetooth scatternet.

several devices to be internet worked over an extended distance. This relationship also allows for a dynamic topology that may change during any given session; as a device moves toward and away from the master device in the network, the topology and therefore the relationship of the devices in the immediate network change.

Mobile routers in a Bluetooth network control the changing network topologies of these networks. The routers also control the flow of data between devices that are capable of supporting a direct link to each other. As devices move about in a random fashion, these networks must be reconfigured on-the-fly to handle the dynamic topology. The routing protocols it employs allow Bluetooth to establish and maintain these shifting networks.

A Bluetooth client may be a laptop, a handheld device (e.g., PDA or custom device such as a barcode scanner), desk top, or any other kind of Bluetooth- enabled device. A Bluetooth client is simply a device with a Bluetooth radio and Bluetooth software module incorporating the Bluetooth protocol stack and interfaces.

4.2.5   Security of Bluetooth

Security for the Bluetooth radio path is depicted in Figure 4.13.

As shown in Figure 4.13, security for Bluetooth is provided on the various wireless links, on the radio paths only. In other words, link encryption and authentication may be provided but true end-to-end security is not possible. In the example provided, security services are provided between the PDA and the printer, between the cell phone and the laptop, and between the laptop and the desktop.

Briefly, the three basic security services defined by the Bluetooth specifications are the following:

Confidentiality. Confidentiality, or privacy, is one security goal of Bluetooth. The intent is to prevent information compromise from eavesdropping (passive attack).

Authentication. A second goal of Bluetooth is the identity verification of communicating devices. This service provides an abort mechanism if a device can not authenticate properly.

Authorization. Third goal of Bluetooth is a security service developed to allow the control of resources.

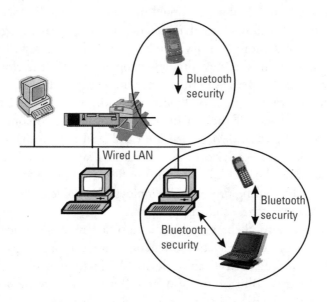

Figure 4.13 Bluetooth air-interface security.

The three security services offered by Bluetooth and details about the modes of security are described in greater detail below. Also worthwhile to note, Bluetooth provides a frequency-hopping scheme with 1,600 hops/second combined with radio link power control (to limit transmit range). These characteristics provide Bluetooth with some additional, albeit small, protection from eavesdropping and malicious access. The frequency-hopping scheme, primarily a technique to avoid interference, makes it slightly more difficult for an adversary to locate the Bluetooth transmission. Using the power control feature appropriately forces any potential adversary to be in relatively close proximity to pose a threat to the Bluetooth network.

4.2.5.1 Security Features of Bluetooth per the Specifications

Bluetooth has three different modes of security. Each Bluetooth device can operate in one mode only at a particular time. The three modes are the following:

Security mode 1: Nonsecure mode;

Security mode 2: Service-level enforced security mode;

Security mode 3: Link-level enforced security mode.

In security mode 1, a device will not initiate any security procedures. In this nonsecure mode, the security functionality (authentication and encryption) is completely bypassed. In effect, the Bluetooth device in mode 1 is in a 'promiscuous' mode that allows other Bluetooth devices to connect to it. This mode is provided for applications for which security is not required, such as exchanging business cards.

In security mode 2, the service-level security mode, security procedures are initiated after channel establishment at the Logical Link Control and Adaptation Protocol (L2CAP) level. L2CAP resides in the data link layer and provides connection-oriented and connectionless data services to upper layers. For this security mode, a security manager (as specified in the Bluetooth architecture) controls access to services and devices. The centralized security manager maintains policies for access control and interfaces with other protocols and device users. Varying security policies and trust levels to restrict access may be defined for applications with different security requirements operating in parallel. Therefore, it is possible to grant access to some services without providing access to other services. Obviously, in this mode, the notion of authorization is introduced– the process of deciding if device A is allowed to have access to service X.

In security mode 3, the link-level security mode, a Bluetooth device initiates security procedures before the channel is established. This is a built-in security mechanism, and it is not aware of any application layer security that may exist. This mode supports authentication (unidirectional or mutual) and encryption. These features are based on a secret link key that is shared by a pair of devices. To generate this key, a pairing procedure is used when the two devices communicate for the first time.

The Bluetooth modes are shown in Figure 4.14.

4.2.5.2 Link Initialization

Prior to communication the Bluetooth devices need to be associated in the link initialization phase. The link key is generated during the initialization phase. Per the Bluetooth specification, two associated devices simultaneously derive link keys during the initializing phase when a user enters an identical personal identification number (PIN) into both devices. The PIN entry, device association, and key derivation are depicted in Figure 4.15. After initialization is complete,

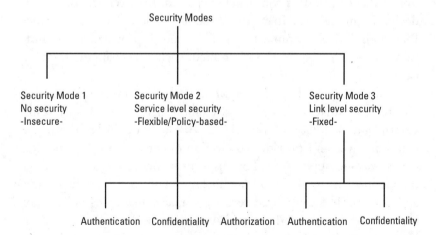

Figure 4.14 Bluetooth security modes.

devices automatically and transparently authenticate and perform encryption of the link. The length of the (PIN) code used in Bluetooth devices can vary between 1 and 16 bytes. The typical 4-digit PIN may be sufficient for some applications, but higher security applications may need longer codes. The PIN code of the device can be fixed, so that it needs to be entered only to the device wishing to connect. Another possibility is that the PIN code must be entered to the both devices during the initialization.

4.2.5.3 Key Management

All security transactions between two or more Bluetooth devices are handled by the link key. The link key is a 128-bit random number. It is used in the authentication process and as a parameter when deriving the encryption key. The lifetime of a link key depends on whether it is a semipermanent or a temporary key. A semipermanent key can be used after the current session is over to authenticate Bluetooth units that share it. A temporary key lasts only until the current session is terminated and it cannot be reused. Temporary keys are commonly used in point-to-multipoint connections, where the same information is transmitted to several recipients.

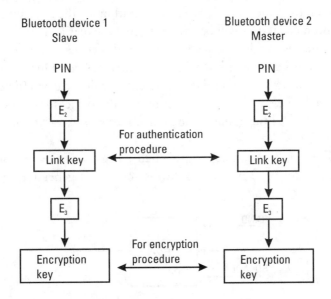

Figure 4.15 Bluetooth key generation.

Link key is termed as a different name depending on the type of application. Link keys can be unit keys, combination keys, master keys, or initialization keys.

The unit key is generated in a single device when it is installed. The combination key is derived from information from two devices and it is generated for each new pair of Bluetooth devices. The master key is a temporary key, which replaces the current link key. It can be used when the master unit wants to transmit information to more than one recipient. The initialization key is used as link key during the initialization process when there are not yet any unit or combination keys. It is used only during the installation.

The initialization key is needed when two devices with no prior engagements need to communicate. During the initialization process, the PIN code is entered to both devices. As shown in Figure 4.16, the initialization key itself is generated by the E22 algorithm, which uses the PIN code, the Bluetooth device address of the claimant device (BD_ADDR) and a 128-bit random number generated by the verifier device as inputs. The resulting 128-bit initialization key is used for key exchange during the generation of a link key. After the key exchange the initialization key is discarded.

The unit key is generated with the key generating algorithm E21 when the Bluetooth device is in operation for the first time. After it has been created, it will be stored in the nonvolatile memory of the device and is rarely changed. Another device can use the other device's unit key as a link key between these devices. During the initialization process, the application decides which party should provide its unit key as the link key. If one of the devices is of restricted memory capabilities (i.e., cannot remember any extra keys), its link key is to be used.

The combination key is generated during the initialization process if the devices have decided to use one. It is generated by both

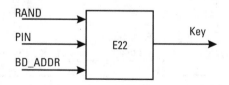

Figure 4.16 Generation of master/initialization key.

devices at the same time. First, both of the units generate a random number. With the key generating algorithm E21, both devices generate a key, combining the random number and their Bluetooth device addresses, as shown in Figure 4.17. After that, the devices exchange securely their random numbers and calculate the combination key to be used between them.

The master key is the only temporary key of the link keys described above. It is generated by the master device by using the key generating algorithm E22 with two 128-bit random numbers. A third random number is then transmitted to the slave and with the key generating algorithm and the current link key an overlay is computed by both the master and the slave. The new link key (the master key) is then sent to the slave, bitwise XORed with the overlay. With this, the slave can calculate the master key. This procedure must be performed with each slave with which the master wants to use the master key.

4.2.5.4 Authentication

The Bluetooth authentication procedure is in the form of a challenge-response scheme. Two devices interacting in an authentication procedure are referred to as the claimant and the verifier. The verifier is the Bluetooth device validating the identity of another device. The claimant is the Bluetooth device attempting to prove its identity. The challenge-response protocol validates devices by verifying the knowledge of a secret key—a Bluetooth link key. The challenge-response verification scheme is depicted conceptually in Figure 4.18. As shown, one of the Bluetooth devices (the claimant) attempts to reach and connect to the other (verifier).

The steps in the authentication process are the following:

Step 1. The claimant transmits its 48-bit address (BD_ADDR) to the verifier.

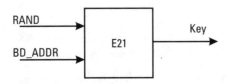

Figure 4.17 Generation of unit/combination key.

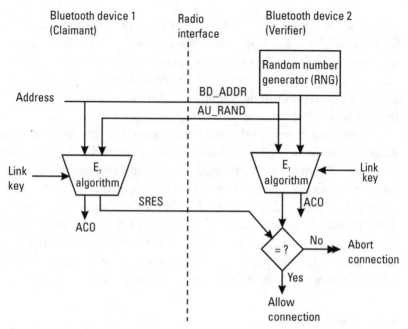

BD_ADDR`–48 bits device address
AU_RAND –128 bits random challenge
ACO –96 bits authenticated cipher offset

Figure 4.18   Bluetooth authentication.

Step 2. The verifier transmits a 128-bit random challenge (AU_RAND) to the claimant.

Step 3. The verifier uses the E_1 algorithm to compute a response using the address (BD_ADDR), link key and the random challenge (AU_RAND) as inputs. The claimant also performs the same computation.

Step 4. The claimant returns the computed response, SRES, to the verifier.

Step 5. The verifier compares the SRES received from the claimant with the SRES that computes.

Step 6. If the two 32-bit SRES values are equal, the verifier will continue connection establishment.

If the authentication fails, a Bluetooth device will wait an interval of time before a new attempt can be made. This time interval will increase exponentially to prevent an adversary from repeated attempts to gain access by defeating the authentication scheme through trial-and-error with different keys. However, it is important to note that this "suspend" technique does not provide security against sophisticated adversaries performing offline attacks to exhaustively search PINs.

Again, the Bluetooth standard allows both unidirectional and mutual- authentication to be performed. The authentication algorithm used for the validation is based on the SAFER+ algorithm. SAFER stands for *secure And fast encryption routine* developed by James Massey. The SAFER algorithms are iterated block ciphers (IBC). In an IBC, the same cryptographic function is applied for a specified number of rounds.

The Bluetooth address is a public parameter that is unique to each device. This address can be obtained by a device enquiry process. The private key, or link key, is a secret key. The link key is derived during initialization, is never disclosed outside the Bluetooth device, and is never transmitted over the air interface. The random challenge, obviously a public parameter is designed to be different on every transaction. The random number is derived from a pseudo-random process within the Bluetooth device. The cryptographic response, SRES, is public as well. With knowledge of the challenge and response parameters, it should be impossible to predict the next challenge or derive the link key.

The parameters used in authentication procedure are summarized in Table 4.2.

4.2.5.5 Data Confidentiality

In addition to the authentication scheme, Bluetooth provides for a confidentiality security service to thwart eavesdropping attempts on the air-interface. Bluetooth encryption is provided to protect the payloads of the packets exchanged between the two Bluetooth devices. The encryption scheme is shown in Figure 4.19.

Bluetooth encryption procedure is based on a stream cipher, E_0. A keystream output is XORed with the payload bits and sent to the receiving device. This key stream is produced using a cryptographic

Table 4.2
Summary of the Authentication Parameters

Parameter	Length	Secrecy Characteristics
Device address	48 bits	Public
Random challenge	128 bits	Public, unpredictable
Authentication response (SRES)	32 bits	Public
Link key	128 bits	Secret

algorithm based on linear feedback shift registers (LFSR). The encryption function takes as inputs, the master identity (BD_ADDR), the random number (EN_RAND), a slot number and an encryption key, which initialize the LFSRs before the transmission of each packet, if encryption is enabled. Since the slot number used in the stream cipher changes with each packet, the ciphering engine is also reinitialized with each packet even though the other variables remain static.

As shown in the figure, the encryption key provided to the encryption algorithm is produced using an internal key generator (KG). This key generator produces stream cipher keys based on the link key, random number (EN_RAND), and the ACO value. The ACO parameter, a 96-bit authenticated cipher offset, is another output produced during the authentication procedure.

The encryption key (K_C) is generated from the current link key. The key size may vary from 8 bits to 128 bits and is negotiated. The negotiation process occurs between master and slave devices. During negotiation, a master device makes a key size suggestion for the slave. In every application, a minimum acceptable key size parameter can be set to prevent a malicious user from driving the key size down to the minimum 8 bits 0 rendering the link totally insecure.

The Bluetooth specification also allows three different encryption modes to support the confidentiality service:

Encryption mode 1: No encryption is performed on any traffic.

Encryption mode 2: Broadcast traffic goes unprotected (not encrypted), but individually addressed traffic is encrypted with the master key.

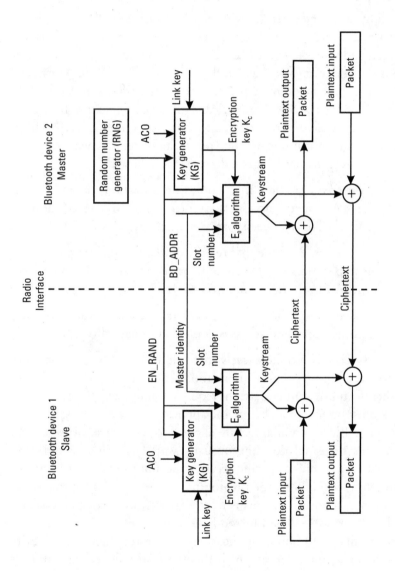

Figure 4.19 Bluetooth encryption/decryption procedures.

Encryption mode 3: All traffic is encrypted with the master key.

4.2.5.6   Trust Levels, Service Levels, and Authorization

In addition to the three security modes, Bluetooth allows two levels of trust and three levels of service security. The two levels of trust are trusted and untrusted. Trusted devices are ones that have a fixed relationship and therefore have full access to all services. Untrusted devices do not maintain a permanent relationship; this results in a restricted service access. For services three levels of security have been defined. These levels are provided so that the requirements for authorization, authentication and encryption can be set independently.

The security levels can be described as follows:

Service level 1: Those that require authorization and authentication. Automatic access is granted only to trusted devices. Untrusted devices need manual authorization.

Service level 2: Those that require authentication only. Access to an application is allowed only after an authentication procedure. Authorization is not necessary.

Service level 3: Those that are open to all devices. Authentication is not required and access is granted automatically.

Associated with these levels are the following security controls to restrict access to services: authorization required (this always includes authentication), authentication required, and encryption required (link must be encrypted before the application can be accessed).

The Bluetooth architecture allows for defining security policies that can set trust relationships in such a way that even trusted devices can get access only to specific services and not to others. It is important to understand that the Bluetooth core protocols can authenticate only devices, not users. This is not to say that user based access control is not possible. The Bluetooth security architecture (through the security manager) allows applications to enforce their own security policies. The link layer, at which Bluetooth specific security controls operate, is transparent to the security controls imposed by application layers. Thus it is possible to enforce user-based authentication and fine-grained access control within the Bluetooth security framework.

4.2.6 Problems in the Security of Bluetooth

There is a problem in the usability of the Bluetooth devices. The use of the PIN code in the initialization process of two Bluetooth devices is tacky. When one has to enter the PIN code twice every time he connects two devices, it gets annoying even with shorter codes. If there is an ad hoc network of Bluetooth devices and every machine is to be initialized separately, it is unbearable, and it does not make upholding the security very easy. The specification makes a suggestion to use application level key agreement software with the longer (up to 16 octets) PIN codes. The PIN code need not be entered physically to each device of the connection, but is exchanged with, for example, Diffie-Hellman key agreement.

The generation of the initialization key may also be of concern. The strength of the initialization key is based purely on the used PIN code. The E22 initialization key generation algorithm derives the key from the PIN code, which is transmitted over the air. The output is highly questionable, as the only secret is the PIN code. When using four-digit PIN codes there are only 10,000 different possibilities. Adding the fact that 50% of used PINs are "0000," the trustworthiness of the initialization key is quite low.

There is also a problem in the unit key scheme. Authentication and encryption are based on the assumption that the link key is the participants' shared secret. All other information used in the procedures is public. Now, suppose that devices A and B use A's unit key as their link key. At the same time (or later on), device C may communicate with device A and use A's unit key as the link key. This means that device B, having obtained A's unit key earlier, can use the unit key with a faked device address to calculate the encryption key and therefore listen to the traffic. It can also authenticate itself to device A as device C and to device C as device A.

The Bluetooth device address, BD_ADDR, which is unique to each and every Bluetooth device, introduces another problem. When a connection is made that a certain Bluetooth device belongs to a certain person, it is easy to track and monitor the behavior of this person. Logs can be made on all Bluetooth transactions and privacy is violated.

Also, there is no end-user authentication. Authentication is only for the devices. A Bluetooth-equipped device can be fraudulently used.

Yet another problem with Bluetooth is the battery draining denial of service scheme, against which it has no protection. If this is going to be a big problem, some countermeasures need to be taken by the Bluetooth SIG.

All in all, there are still several problems in the security of Bluetooth. It seems to be adequate for smaller applications, but any sensitive or otherwise problematic data should not be transmitted via Bluetooth.

Reference

[1]   Fluhrer, S., I. Mantin, and A. Shamir, "Weaknesses of the Key Scheduling Algorithm of RC4," *Proc. of 8th Annual Workshop on Selected Areas in Cryptography, LNCS 2259*, Springer-Verlag, August 2001, pp. 1–24.

5

Security in 2G Systems

With the advancement in technology, people wanted to communicate from anywhere, at anytime. A first generation analog system in mobile communication incorporates mobility in the system and is used for voice communication only. The second generation system moves to the digital era from the analog system, still being used for voice communication but also including some sort of data communications. When mobile communication came into being, security was not considered, but this was incorporated in the second generation systems.

Global system for mobile (GSM) communication provides true mobile communications with support for voice and data services and allows for worldwide connectivity and international mobility with unique subscriber address. CDPD was then developed to transfer digital data over the first generation analog standard AMPS in America. In Japan, I-mode is widely used for mobile communication. This is a higher layer standard running on top of PDC. In this chapter, the security issues, and the weaknesses of the security consideration of these systems, are discussed.

5.1 GSM System

5.1.1 Introduction

During the early 1980s, analog cellular telephone systems were experiencing particularly rapid growth in Scandinavia and the United Kingdom, as well as in France and Germany. Each country developed its own system, each incompatible with everyone else's system in equipment and operation. This was an undesirable situation, because not only was the mobile equipment limited to operation within national boundaries in an increasingly unified Europe, but there was also a very limited market for each type of equipment, so economies of scale and the subsequent savings could not be realized.

The Europeans realized this early on, and, so in 1982, the Conference of European Posts and Telegraphs (CEPT) formed a study group called the Groupe Spécial Mobile (GSM) to study and develop a pan-European public land mobile system. The proposed system had to meet certain criteria:

- Subjective speech quality;
- Low terminal and service cost;
- Support for international roaming;
- Ability to support handheld terminals;
- Support for range of new services and facilities;
- Spectral efficiency;
- ISDN compatibility.

In 1989, GSM responsibility was transferred to the European Telecommunication Standards Institute (ETSI), and, in 1990, Phase-I of the GSM specifications were published. Commercial service began in mid-1991. Although standardized in Europe, GSM is not only a European standard, but exists on every continent, and the acronym GSM now aptly stands for global system for mobile communications.

GSM now boasts 863.6 million subscribers in 197 countries and constitutes 70% of the world's wireless market. Some of the more widely known problems with GSM's analog counterparts included the possibility of phone fraud by cloning phones and making calls at the owner's expense, and the possibility of someone intercepting phone calls over the air and eavesdropping on the caller's discussion. The GSM system was developed to correct these problems by implementing strong authentication between the mobile station (MS) and the mobile switching center (MSC), as well as implementing strong data encryption for the over-the-air transmission channel between the MS and the base tranceiver station (BTS).

The GSM specifications were designed by the GSM consortium in secrecy and were distributed strictly on a need-to-know basis to hardware and software manufacturers and to GSM network operators. The GSM security algorithms were developed privately by the Security Algorithm Group of Experts (SAGE), comprised of leading cryptographers. The specifications were not intended as public information, but in the mid-1990s, information about the algorithms slowly leaked into the public domain.

5.1.2 Architecture of the GSM Network

A GSM network is composed of several functional entities, with specified functions and interfaces. Figure 5.1 shows the layout of a generic GSM network. The GSM network can be divided into three broad parts, consisting of the MS carried by the subscriber, the BSS, which controls the radio link with the MS, and the network subsystem, the main part of which is the MSC. The MSC performs the switching of calls between mobile users, and between mobile and fixed network users. The MSC also handles the mobility management operations. Not shown is the operations and maintenance center (OMC), which oversees the proper operation and setup of the network. The MS and the base station subsystem (BSS) communicate across the U_m interface, also known as the air interface or radio link. The BSS communicates with the MSC across the A interface.

5.1.2.1 Mobile Station (MS)

MS consists of the mobile equipment (the terminal) and a smart card called the subscriber identity module (SIM). The SIM provides personal mobility, so that the user can have access to subscribed services irrespective of a specific terminal. By inserting the SIM card into

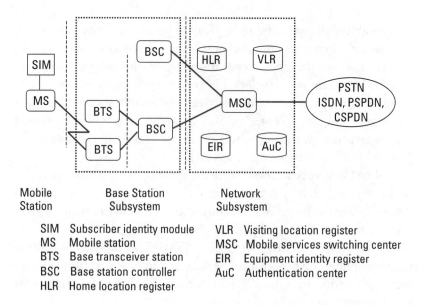

Figure 5.1 General architecture of a GSM network.

another GSM terminal, the user is able to receive calls at that terminal, make calls from that terminal, and receive other subscribed services.

The mobile equipment is uniquely identified by the international mobile equipment identity (IMEI). The SIM card contains the international mobile subscriber identity (IMSI) used to identify the subscriber to the system, a secret key for authentication, and other information. The IMEI and the IMSI are independent, thereby allowing personal mobility. The SIM card may be protected against unauthorized use by a password or personal identity number.

5.1.2.2 Base Station Subsystem (BSS)

The base station subsystem is composed of two parts: the base transceiver station (BTS) and the base station controller (BSC). These communicate across the standardized Abis interface, allowing (as in the rest of the system) operation between components made by different suppliers. BTS houses the radio transceivers that define a cell and handles the radio-link protocols with the MS. The BSC manages the radio resources for one or more BTSs. It handles radio-channel setup, frequency hopping and handovers. The BSC is the connection between the MS and the MSC.

5.1.2.3 Network Subsystem

The central component of the network subsystem is the MSC. It acts like a normal switching node of the PSTN or ISDN, and additionally provides all the functionality needed to handle a mobile subscriber, such as registration, authentication, location updating, handovers, and call routing to a roaming subscriber. These services are provided in conjunction with several functional entities, which together form the network subsystem. The MSC provides the connection to the fixed networks (such as the PSTN or ISDN). Signaling between functional entities in the network subsystem uses signaling system number 7 (SS7), used for trunk signaling in ISDN and widely used in current public networks.

The home location register (HLR) and visitor location register (VLR), together with the MSC, provide the call-routing and roaming capabilities of GSM. The HLR contains all the administrative information of each subscriber registered in the corresponding GSM network, along with the current location of the MS. The location of the mobile station is typically in the form of the signaling address of the VLR associated with the MS. There is logically one HLR per GSM network, although it may be implemented as a distributed database.

The VLR contains selected administrative information from the HLR, necessary for call control and provision of the subscribed services, for each mobile currently located in the geographical area controlled by the VLR. Although each functional entity can be implemented as an independent unit, all manufacturers of switching equipment to date implement the VLR together with the MSC, so that the geographical area controlled by the MSC corresponds to that controlled by the VLR, thus simplifying the signaling required. Note that the MSC contains no information about particular mobile stations—this information is stored in the location registers.

The other two registers are used for authentication and security purposes. The equipment identity register (EIR) is a database of blacklisted cell phones. It represents a list of the IMEI of cell phones reported stolen and subsequently placed on the EIR. Every cell phone is identified by its IMEI. When a cell phone is connected to the network, the handset's IMEI is read by the network. If the handset is listed as stolen on the blacklist, it can be disabled electronically and will then be unusable on many GSM cellular networks around the world.

The authentication center (AuC) is a protected database that stores a copy of the secret key stored in each subscriber's SIM card, which is used for authentication and encryption over the radio channel.

5.1.3 GSM Security Features

The security goal for GSM system was to make the system as secure as the public switched telephone network (PSTN) and to prevent phone cloning. The use of air-interface as the transmission media allows a number of potential threats of eavesdropping the transmissions. The security of GSM system was built to protect the air-interface communication. In fact there was no attempt to provide security on the fixed network part of GSM. In this section brief overview of the security features of GSM system is given.

Security in GSM consists of the following aspects:

- Subscriber identity authentication;
- User and signaling data confidentiality;
- Subscriber identity confidentiality.

The subscriber is uniquely identified by the IMSI. This information, along with the individual subscriber authentication key (K_i), constitutes sensitive identification credentials. The design of the GSM authentication and encryption schemes is such that this sensitive information is never transmitted over the radio channel. Rather, a challenge-response mechanism is used to perform authentication. The actual conversations are encrypted using a temporary, randomly generated ciphering key (K_c). The MS identifies itself by means of the temporary mobile subscriber identity (TMSI), which is issued by the network and may be changed periodically (i.e.. during handoffs) for additional security.

The security mechanisms of GSM are implemented in three different system elements: the SIM, the GSM handset or MS, and the

GSM network. The SIM contains the IMSI, the individual subscriber authentication key (K_i), the ciphering key generating algorithm (A8), the authentication algorithm (A3), as well as a personal identification number (PIN). The GSM handset, MS contains the ciphering algorithm (A5). All three algorithms (A3, A5, and A8) are present in the GSM network as well. The authentication center (AuC), part of the operation and maintenance subsystem (OMS) of the GSM network, consists of a database of identification and authentication information for subscribers. The IMSI, and the individual subscriber authentication key (K_i) for each user are stored in the AuC, as well as the A3 and A8 algorithms. In order for the authentication and security mechanisms to function, all three elements (SIM, handset, and GSM network) are required.

Figure 5.2 demonstrates the distribution of security information among the three system elements: the SIM, the MS, and the GSM network. Within the GSM network, the security information is further distributed among the authentication center (AuC), the home location register (HLR), and the visitor location register (VLR). The AuC is responsible for generating the sets of triplets (*RAND, SRES, K_c*) which are stored in the HLR and VLR for subsequent use in the authentication and encryption processes.

5.1.3.1 Subscriber Identity Authentication

This subscriber identity authentication service is the core of the GSM security system. It is used to enable the fixed network to authenticate the identity of mobile subscriber, and to establish and manage the

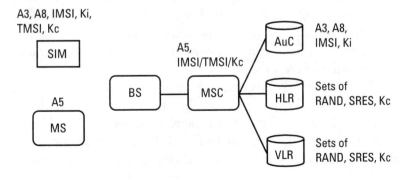

Figure 5.2 Distribution of security features in the GSM network.

encryption keys needed to provide the confidentiality services. The service must be supported by all networks and mobiles, although the frequency of application is at the discretion of the network.

Authentication is initiated by the fixed network, and is based upon a simple challenge-response protocol. When a mobile station needs to authenticate to a serving network, one of the following situation may occur:

Case 1: The visiting cell belongs to a network which the MS has not visited in the (recent) past. In this case, it presents its IMSI to the serving network. The serving networks MSC finds out the visiting MS's home network and asks the HLR to send an authentication vector, which is stored in the serving networks VLR together with the IMSI of the MS.

Case 2: The visiting cell belongs to the home network to which the MS belongs or to a network to which the MS has already authenticated in the (recent) past. If the authentication vector of the mobile station is still available in the VLR and there are some triplets left unused, then the HLR of the visiting MS does not need to be contacted.

In both cases, an unused random challenge $RAND$ is sent to the MS. The MS computes a response $SRES$ to $RAND$ using a one-way function A3 under control of a subscriber authentication key, K_i. The key K_i is unique to the subscriber, and is shared only by the subscriber and an authentication center (AuC) which serves the subscriber's home network. The value $SRES$ computed by the MS is signaled to the network, where it is compared with a precomputed value. If the two values of $SRES$ agree, the mobile subscriber has been authenticated, and the call is allowed to proceed. If the values are different, then access is denied. Figure 5.3 illustrates the GSM subscriber identity authentication scheme.

In case one, the visiting MS attempts to access the visiting domain network for the first time when it sends its IMSI to the visited network. BSC of the visited network sends this IMSI to the home network and home networks HLR sends a set of triplets ($RAND$, $SRES$, K_c) back to the visited network. The visited network uses these triplets to authenticate the visiting MS.

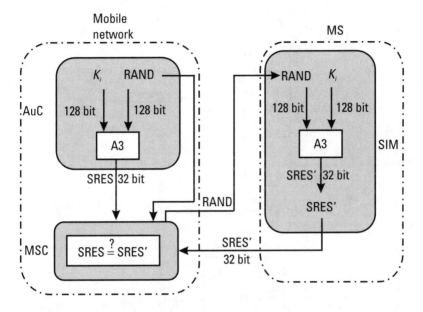

Figure 5.3 GSM subscriber identity authentication scheme.

A3 algorithm takes the random challenge *RAND* and the secret key K_i and generates *SRES* output (Figure 5.4). Both *RAND* and K_i are 128 bits long and output *SRES* is 32 bits long.

The same mechanism is also used to establish a cipher key K_c for encrypting user and signaling data on the radio path. A8 algorithm generates the session key K_c from the random challenge *RAND* and the secret key K_i. A8 algorithm takes these two 128-bit inputs and generates a 64-bit output from them. This output is the 64-bit session key K_c (Figure 5.5). The BTS receives the same K_c from the MSC. AuC is able to generate the K_c, because the HLR knows both the *RAND* (the HLR generated it) and the secret key K_i, which it holds for all the GSM subscribers of this network operator. One session key, K_c, is used until the MSC decides to authenticate the MS again, which might be a matter of days.

The key K_c is computed by the MS using a one-way function A8, again under control of the subscriber authentication key K_i, and is precomputed for the network by AuC that serves the subscriber's home network. Thus at the end of a successful authentication exchange, both parties possess a fresh cipher key K_c.

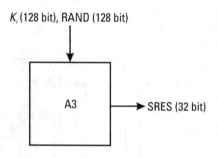

Figure 5.4 SRES calculation.

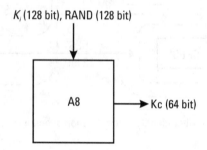

Figure 5.5 Session key, K_C calculation.

The precomputed triplets (*RAND, SRES, K_c*), held by the fixed networks for a particular subscriber, is passed from the home network's authentication center to visited networks upon demand. The challenges are used just once. Thus the authentication center never sends the same triple to two distinct networks, and a network never re-uses a challenge.

COMP128, which combines A3 and A8 into a single algorithm, generates both the *SRES* response and the session key, K_c on one run. COMP128 also takes the 128 bit *RAND* and 128-bit long K_i as input, but it generates 128 bits of output instead of 32 bit *SRES*. The first 32 bits of 128 bit output is the *SRES* response. The last 54 bits of the COMP128 output form the session key, K_c, until the MS is authenticated again. (See Figure 5.6.) Ten zero-bits are appended to the key generated by the COMP128 algorithm. Thus, the session key K_c becomes a key of 64 bits with the last ten bits zeroed out. But this effectively reduces the keyspace from 64 to 54 bits.

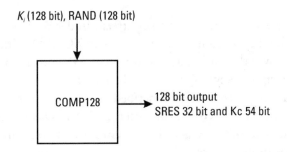

K_i (128 bit), RAND (128 bit)

COMP128

128 bit output
SRES 32 bit and Kc 54 bit

Figure 5.6 COMP128 calculation.

Either COMP128 algorithm or both the A3 and A8 algorithm are stored in the SIM in order to prevent it from tempering. The operator can choose the algorithms independently from hardware manufacturers and other network operators. The authentication works in other countries as well, because the local network asks the HLR of the subscriber's home network for the triplets (*RAND*, *SRES*, K_c). Thus the local network does not have to know anything about COMP128 or A3 and A8 algorithms.

5.1.3.2   User and Signaling Data Confidentiality

This service consists of three elements:

1. User data confidentiality and signaling information on physical connections;
2. Connectionless user data confidentiality;
3. Signaling information element confidentiality.

The first element provides for privacy of all user generated data, both voice and nonvoice, transferred over the radio path on traffic channels. The second element provides for privacy of user data transferred in packet mode over the radio path on a dedicated signaling channel, while the third element provides for privacy of certain user related signaling elements transferred over the radio path on dedicated signaling channel.

All of these elements of service are provided using the same encryption mechanism, and must be supported and used by all networks and mobiles. Encryption is achieved by means of a ciphering algorithm A5 which produces a key stream under control of a cipher key K_c. This key stream is then bit-by-bit XORED with the data transferred over the radio path between the MS and the base station (BS). The cipher key, K_c is established at the MS as part of the authentication procedure, as described in the last section, and BS received it from the fixed network after the MS has been identified. It is essential that the MS and BS synchronize the starting of their cipher algorithms (Figure 5.7).

Synchronization of the ciphering key stream is maintained by using the TDMA frame structure of the radio subsystem. The TDMA frame number is used as a message key for the cipher algorithm A5, and the algorithm produces a synchronized key stream for enciphering and deciphering the data bits in the frame. For each frame, a total of 114 bits are produced for enciphering/deciphering data transferred from the MS to the BS, and an additional 114 bits are produced for deciphering/enciphering data received at the MS from the BS. A frame lasts for 4.6 ms, so that the cipher has to produce the 228 bits in this time.

There are two currently used versions of A5: A5/1 is the stronger, export-limited version, and A5/2 is the weak version that has no export limitation. An additional new version which is standardized but not yet used in GSM networks is A5/3. It was recently chosen and is based on the blockcipher KASUMI.

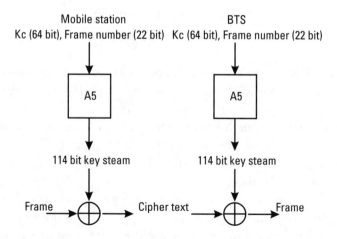

Figure 5.7 Key stream generation and frame encryption/decryption.

5.1.3.3 Subscriber Identity Confidentiality

This service allows mobile subscribers to originate calls, and update their location without revealing the IMSI to an eavesdropper on the radio path. It thus prevents location tracing of individual mobile subscribers by listening to the signaling exchanges on the radio path. All mobiles and networks must be capable of supporting the service, but its use is not mandatory.

In order to provide the subscriber identity confidentiality service, it is necessary to ensure that the IMSI, or any information which allows an eavesdropper to derive the IMSI, is not transmitted in clear in any signaling message on the radio path. The mechanism used to provide this service is based on the use of a temporary mobile subscriber identity (TMSI), which is securely updated after each successful access to the system. Thus, in principle, the IMSI need only be transmitted in clear over the radio path at registration. In addition, the signaling elements which convey information about the IMSI are enciphered.

The TMSI updating mechanism functions in the following manner: For simplicity, assume the MS has been allocated a TMSI, denoted by $TMSI_o$, and the network knows the association between $TMSI_o$ and the subscriber's IMSI. The MS identifies itself to the network by sending $TMSI_o$. Immediately after authentication (if this takes place), the network generates a new TMSI, denoted $TMSI_n$, and sends this to the MS encrypted under the cipher key K_c as described in the last section. Upon receipt of the message, the MS deciphers and replaces $TMSI_o$ by $TMSI_n$.

5.1.4 Attacks on GSM Security

In GSM, all security algorithms were originally kept secret. However, the information was later leaked and exposed to the public, after which the vulnerabilities of the GSM system were exposed and led to an attack on GSM security.

5.1.4.1 Microwave Links

The fact that the BS to BSC link is, in many cases, a point-to-point microwave link, makes it a security hole in the GSM system. This link can be eavesdropped upon as data is, at this point, generally unencrypted. When the GSM was first designed, it was expected that the BS to BSC link would be across fixed links and, therefore, that encryption would not be required. To protect data in microwave link some operators have implemented lower layer bulk encryption.

5.1.4.2 Attacks on the Algorithm A3/8

The Smart Card Developer Association and the ISAAC security research group discovered a flaw in the COMP128 algorithm that effectively enabled them to retrieve the secret key K_i from the SIM in April, 1998 [1]. They proved that if approximately 160,000 chosen *RAND-SRES* pairs could be collected, it was possible to obtain complete knowledge of secret key K_i. The quickest method of attack would be to steal the user's mobile phone, remove the SIM and connect it to a phone emulator that can be used to send 160,000 chosen *RAND* to SIM and receive the *SRES*. SIM tends to have relatively slower clock speeds, and therefore it can take up to 10 hours to obtain 160,000 pairs (with faster SIM, it would take 2 and a half hours). Another possibility is to use a false BS to send the *RAND* over the air interface. Because the rate at which pairs can be collected is slower, it would take a number of days; however the attacker does not need physical possession of SIM.

The security of whole GSM security model is based on the secret key K_i. If this key is compromised then the security of the system for a subscriber is totally lost. Once the attacker is able to retrieve the key, he can not only eavesdrop on the subscriber's calls, but also masquerade as the original subscriber and run calls on the original subscriber's bill.

5.1.4.3 Attacks on A5 Algorithm

If the attacker obtains knowledge of session key K_c, he or she is then capable of finding the key stream used for encryption/decryption data, and the number of frames that are encrypted. This is possible since the key stream is generated from the session key K_c, and while K_c in practice, does not change during the call and may be used for days, the frame numbers are generated implicitly.

A real-time brute force attack against the GSM security protocols is not feasible. Though the length of the key K_c is 64 bits, the least significant bits are always zero the key space reduces to 2^{54} keys. Therefore, too much time is needed for it to be feasible for eavesdropping on GSM calls in real time. Even though the real time eavesdropping seems to be infeasible since it must need time for K_c cryptanalysis; attacker can still record the data frames and decrypt later after a successful brute force attack finding the session key.

In addition to finding the session key K_c by brute force attack, new methods are developing to break A5 in less time, which makes possible real-time eavesdropping. The "divide and conquer" attack initiated by J. Golic can reduce the complexity to $2^{40.16}$ under the assumption of knowing plaintext and trying to determine the initial states of the LFSRs from a known key stream sequence [2]. Alex Biryukov, Adi Shamir, and David Wagner [3] proposed cryptanalytic attacks on A5/1, in which a single PC can extract the session key K_c in real time from a small amount of generated output. The technique used is known as "time-memory trade-off." In a preprocessing phase, a large database of algorithm states and related key stream sequences are created. In the attack phase, the data base is searched for a match with subsequences of the known key stream. If a match is found, it is highly probable that the database will give the correct algorithm state. From there, it is simple to compute the session key K_c and decipher the rest of the call.

5.1.4.4   Side Channel Attacks

Side channel attacks are indirect approach cryptanalytic attacks that determine the relationship between the input–output information that leaks from the side channels during the computation such as power consumption, timing of operations, and the like. A strong algorithm against side channel attack must have side channel information independent from input, output and secret information, such as keys. However, if inadequate implementation is used, due to the device resource or cost limitation, some resistance to some side channel attacks may be possible, and some statistical dependent susceptible to attack may still remain.

With physical access to the SIM card, it is still possible to extract secret key K_i by a side channel attack method called "partition attack" developed by IBM researchers. This attack can be successfully implemented in cases where large table lookups are used or where countermeasures against differential side channel analysis have not been properly applied. COMP128 algorithm on the SIM cards, which use a large table lookup, is found to be broken by partition attack with less then 1,000 invocations with random inputs, or 255 chosen inputs or only 8 adaptively chosen inputs. This can extract the secret key within a minute [4].

If the GSM security model is broken on many levels, it is vulnerable to numerous attacks targeted at different parts of an operator's network. Assuming that the security algorithms were not broken, the GSM architecture will still be vulnerable to attacks targeting the operator's backbone network or HLR and to various social engineering scenarios in which, for instance, the attacker bribes an employee of the operator. Furthermore, the secretly designed security algorithms incorporated into the GSM system have been proven faulty. So we see that although the GSM standard was supposed to correct the problems of phone fraud and call interception found in the analog mobile phone systems by using strong crypto for MS authentication and over-the-air traffic encryption, these promises were not kept.

5.2 I-mode Introduction

5.2.1 Introduction

I-mode is a service and a system that enables WEB browsing and e-mail over cellular networks. NTT DoCoMo of Japan started this service in Japan in February 1999. Since then, I-mode has prevailed and became a massive service with about 40 million Japanese users, and is also expanding overseas, gradually increasing its users in Europe and Asia. In Germany, I-mode service started in March 2002 by E-plus, in the Netherlands by KPN Mobile and in France by Bouygues Telecom. I-mode has now become a global standard as an application service using cellular network.

I-mode is basically independent of bearers and aims at providing common contents to users. Therefore, there is basically no difference between I-mode as a service provided over the second generation telecommunication system and that over the third generation system. However, I-mode as a system realized on each telecommunication system differs according to the generation and to the development background of each telecommunication system. There are realization techniques and "know-hows" for each system. This section provides an overview of I-mode security structured on the second generation PDC-P system.

The phrase "security of I-mode" usually suggests security of the I-mode system as a whole, including security against attacks on servers and nodes, and security against spam mail. However, this document gives an overview of I-mode system mechanism that ensures security of data transmitted over communication channels.

5.2.2 I-mode System Overview

As mentioned previously, when I-mode service started in Japan, the system was structured based on PDC-P system, which is one of the second generation telecommunication systems. PDC-P system is a packet data telecommunication system that realizes full duplex communication. To increase throughput at a low data rate, it is necessary to minimize the amount of overhead of application data. Therefore, a simplified protocol unique to NTT DoCoMo was adopted for transport layer and presentation layer.

As shown in Figure 5.8, security was ensured by introducing scrambling in radio interface, dedicated line for cable interface and SSL in Internet when required. However, to realize services such as transactional systems for electronic commerce, a secure end-to-end communication channel was required. Therefore, SSL (secure socket

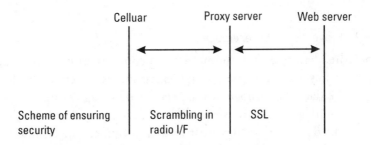

Figure 5.8 Discontinuously secure platform.

layer) [5, 6] was introduced in January 2001 to structure platform that can ensure security of application data.

5.2.3   SSL Overview

SSL is a security protocol for the application data that is advocated by Netscape Communications. As a draft version of SSL is available on the Web, SSL has become a de facto standard as a security protocol. It is located immediately above the transport layer and is a protocol that ensures security against attacks such as *impersonating, eavesdropping, and tampering* in the upper layers during data communications.

The cryptosystem in SSL is generally referred to as a hybrid system. In a hybrid system, application data itself is encrypted using common key cryptosystem and the source data that generates the common key is encrypted using public key cryptosystem.

5.2.3.1   Prevention of Impersonating

Before starting encrypted communication with the desired counterpart, the certificate of the counterpart is verified and its identity is confirmed to prevent impersonation from a third party using fake domains. The data format of a certificate is specified by RFC3280 and has the structure shown in Figure 5.9. Public key cryptosystem is used for verification of a certificate and by verifying with a public key the signature part that is created by a private key, the issuer of the certificate is determined. In addition, one-way hash function is used to compute the message digest of the certificate data and is compared with the signature part to identify whether the certificate has been tampered with.

5.2.3.2   Prevention of Eavesdropping

When the identity of the communicating counterpart is confirmed, a common key used for encryption/decryption is generated in both sides. The data that follows will be encrypted using asymmetric key cryptosystem and the transmitted data cannot be decrypted by a vicious third party that may exist in a communication channel if a common key cannot be obtained. A simplified figure of common key generation procedure is shown in Figure 5.10.

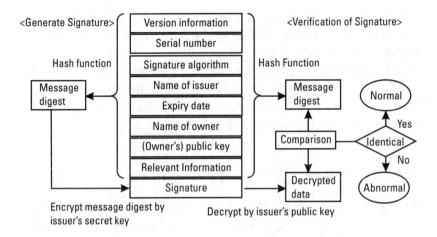

Figure 5.9 X.509 certificate.

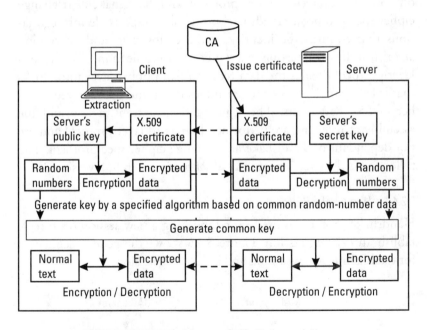

Figure 5.10 Procedures for generating common key.

5.2.3.3 Prevention of Tampering

A third party existing in a communication channel cannot decrypt the content of an encrypted data but can modify the data itself. It is not possible to prevent tampering of the data but by attaching message authentication code (MAC) to the transmitted data it is possible to detect whether the data has been tampered with. MAC is a value that is computed using hash function from the application data and a MAC key that is held in secret by communicating parties. A third party cannot learn the MAC key from transmitted data. The scheme of computing MAC is shown in Figure 5.11.

5.2.4　Protocol Stack

As shown in Figure 5.12, SSL is located as the upper layer of TL and encrypts the data on AL. SSL is performed by request from AL. AL can also use TL directly without using SSL.

　　　The SSL protocol itself is divided into two layers immediately on top of TL: record layer protocol and handshake/alert/change cipher spec protocol. Handshake protocol is responsible for negotiations that confirm the identity of the counterpart and encryption algorithm before starting encrypted communication. The record layer protocol divides the data received from upper layer into blocks that is less than 2^{14} bytes and assembles them into the required number of SSL data units. The segmentation at transmission and the assembly at receiving are performed by record layer protocol and are not dependant on handshake/alert/change cipher spec protocol. The following is the overview of individual protocols in SSL version 3.

5.2.4.1　Handshake Protocol

Handshake protocol is used for establishing a new session or to reestablish an existing session. Figure 5.13 shows the general sequences for establishing a new session and an existing session.

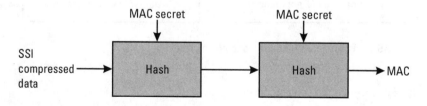

Figure 5.11　Scheme of computing MAC.

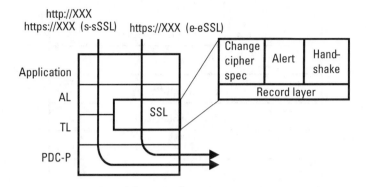

Figure 5.12 Protocol stack.

Client Hello

Client Hello message is sent from a client in the following three cases: when it initially connects to server, when receiving *Hello Request* from server, and when changing, for instance, encryption parameters of an existing connection. Client sends to a server encryption algorithms and compression algorithms that the client supports by *Client Hello* message and waits for the server to send back *Server Hello* message. If reestablishing an existing session, handshake protocol process can be reduced by designating the previous session ID in the body part of this message.

Server Hello

The server receiving *Client Hello* message sends back *Server Hello* message to the client. *Server Hello* message designates an encryption algorithm and a compression algorithm to be used from ones that are suggested from the client. Session ID is held in this message to determine whether or not a new session is established.

Server Certificate

The server sends a certificate that shows its identity to the client using *Server Certificate* message. SSL assumes that there are certification authority (*CA*) structured in layers and the format of a certificate is the format specified by ITU-T recommendation X.509 (almost equal to RFC3280) [7]. The body part of this message is simple and consists of a list of certificates from the server itself to the top-level *CA*.

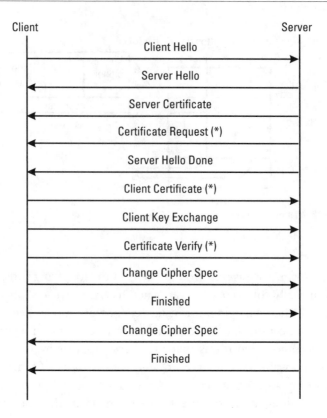

Client   Server

Client Hello

Server Hello

Server Certificate

Certificate Request (*)

Server Hello Done

Client Certificate (*)

Client Key Exchange

Certificate Verify (*)

Change Cipher Spec

Finished

Change Cipher Spec

Finished

Figure 5.13 SSLv3 sequence.

Certificate Request

A server can request a client to show its certificate. *Certificate Request* message is used for such a case. By using this message, the server suggests the client the desired certificate types and *CA*'s names according to priority.

Server Hello Done

When completing process up to *Certificate Request* transmission, a server sends *Server Hello Done* message to inform the client that process in the server side has completed, then waits for client's response.

Client Certificate (Optional)

When there is a request from a server to show a certificate (*Certificate Request* message), this is the first message sent to a server from a client after the client receives *Server Hello Done* message. The message structure is the same as *Server Certificate* message.

Client Key Exchange

Client Key Exchange message has a premaster secret for generating a master secret that is 48 bytes long and must be secretly shared by a server and a client. The master secret can be computed using premaster secret. The body part of *Client Key Exchange* message is encrypted by the algorithm designated by the server certificate.

Certificate Verify (Optional)

Certificate Verify message is a message that is sent to support a server to authenticate the certificate of a client. This message will not be sent if the algorithm designated by a certificate does not support cryptography using a private key. The data in the body part of this message is the message digest encrypted by the private key of a client. The message digest is extracted from data already known to both the server and the client. MD5 and SHA are used as message digest algorithms, so even when a defect occurs in one algorithm, the message will be protected by the other algorithm.

Finished

Finished message is sent to show that handshake protocol has completed successfully. Both the client and server transmit this message when the message transmission is completed and the authentication of the counterpart performs normally. At this point, the client and the server have completed a negotiation regarding compression algorithm and encryption algorithm to be used and has exchanged *change/cipher/spec protocol*. Thus, the data that follows (including *Finished* message) will be transmitted/received after it is processed by the new algorithm. The body part of *Finished* message includes the data that assures the identity of the server and the client. Therefore, the process, after receiving *Finished* message needs to confirm that the data is correct.

5.2.4.2 Alert Protocol

Alert protocol is used to inform the counterpart that an error has occurred during protocol processing. There are two types according to the error level: warning and fatal. If a fatal-level error occurs, the connection will be terminated immediately and the session will be set to a status that does not allow new connection This will not, however, affect other connection(s) that already exist in the same session.

5.2.4.3   Change/Cipher/Spec Protocol

Change/cipher/spec protocol is used to inform the communicating counterpart that the encryption algorithm used will be changed. After the negotiation regarding encryption algorithm used in handshake protocol is completed, this message declares the chosen algorithm. This message is usually sent from both the client and the server at a specific timing. However, if this message is received at an unexpected timing, it will be processed as an error.

5.2.4.4   Application Data Protocol

This protocol is used to exchange actual application data after completing handshake protocol.

5.2.5   HTTP Tunneling Protocol

As mentioned earlier, SSL communication service was provided between proxy server and Web server (server-to-server SSL, hereafter s-sSSL) (Figure 5.7) when I-mode was initially developed. Thus, it was necessary to consider coexistence with SSL communication service between cellular and Web server (end-to-end SSL, hereafter e-eSSL). This section gives an overview of an example of system structure to realize coexistence of both services.

I-mode is an Internet connection via a proxy server. Therefore, tunneling protocol [8] is adopted for realization of e-eSSL as it is necessary to perform communication transparent to proxy servers. If the scheme of connected URL is "https," tunneling request with connected URL information attached is sent to proxy server before an HTTP request is sent to the connected address. The proxy server will determine whether to allow tunneling or not according to URL information and sends the result back to the cellular.

5.2.5.1   s-sSSL

The proxy server has a database of URLs of information contents providers that demands s-sSSL service and if the connected URL of tunneling request is an s-sSSL service-applicable server, it rejects tunneling request to the mobile. If the tunneling request is rejected, the cellular realizes that connection, uses s-sSSL, and will send HTTP request to proxy server without using SSL layer. The proxy server that received HTTP request determines that it is an SSL communication from URL scheme "https" and sends HTTP request to Web server after establishing SSL session. Figure 5.14 shows the sequence of using SSLv3 during s-sSSL connection.

5.2.5.2 e-eSSL

A proxy server will accept a tunneling request if the connected URL does not exist in the database. When a tunneling request is accepted,

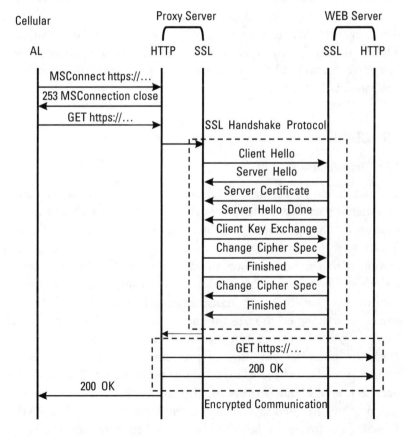

Figure 5.14 s-sSSL sequence.

the mobile identifies that connection system is e-eSSL and sends HTTP request to WEB server after establishing SSL session. After tunneling request is accepted, the communication will be transparent to proxy server and a continuous and secure communication channel will be provided end-to-end. Figure 5.15 shows the sequence of using SSLv3 during e-eSSL connection.

5.2.6 Postscript

This section describes data security in I-mode system. As previously mentioned, I-mode Web browsing has realized a secure communication platform by introducing SSL. However, there are issues, including handling of SSL client authentication, two-way authentication function in application level, and encryption of e-mails such as S/MIME. One method of dealing with two-way authentication function and S/MIME is to equip each mobile with X.509 certificate. There are also many operational issues for each operators, such as key generation, PKCS#10 application procedure, and certificate download I/F.

5.3 CDPD

5.3.1 Introduction

Cellular digital packet data (CDPD) is an attempt to provide access to a digital packet-switched network by using part of the infrastructure of an analog telephone cellular network such as AMPS. The most straightforward approach in achieving this goal is to connect a computer to a standard modem and then connecting it to an analog cell phone. However, this approach does not offer the reliability and security necessary for data transmission, since analog cellular networks were designed to transmit voice and not data, which is much more fragile to error transmissions. Moreover, analog cellular networks do not offer any kind of protection against eavesdropping. CDPD provides a cheap solution to this problem. CDPD is a mobile data technology that permits subordinate packet data operation on the spectrum assigned to a telephone cellular network, such as AMPS. It was first proposed by IBM as a packet-switching overlay to the existing analog cellular network and frequencies. In 1993, a

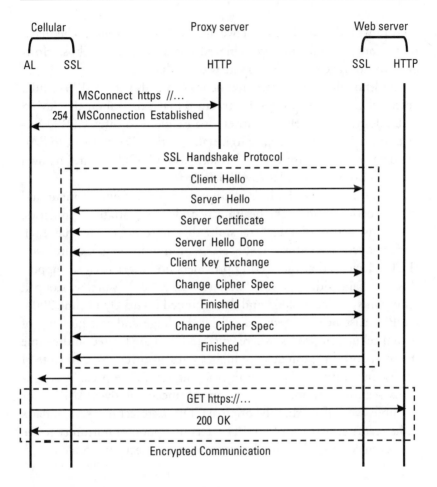

Figure 5.15 e-eSSL sequence.

consortium of American cellular carriers (Ameritech, Bell Atlantic, Contel, GTE, Mcaw, NYNEX, Airtouch, and Southwest Bell) developed the original CDPD specifications. By 2000, CDPD use was well established.

The basic idea of CDPD is to send data in digital format during idle times, that is, when voice calls are not being made. It uses the same 30 KHz channels at 800 MHz as the normal analog cellular system. The modulation used in the RF channel was GMSK with Reed-Solomon (63, 39) forward error correction. The advantage of using the same frequency of cellular systems is that, potentially,

CDPD would have the same coverage as cellular systems. The maximum data transmission rate achieved was around 19.2 Kbps. However, the average transmission rate was much lower.

Although CDPD shares frequency channels with AMPS cellular voice calls, it has its own infrastructure that exists upon the AMPS technology. Thus, cellular carriers who choose to offer CDPD services to their subscribers have to install additional equipment to handle data separately from AMPS voice. Also, CDPD requires its own modems not using regular AMPS handsets.

CDPD was, and still is, mostly used for law enforcement and public services, and in the health care and transportation industries, where just getting the data transmitted is more important than high performance. Recently, the Federal Communications Commission (FCC) decided that the CDPD providers no longer need to support this low-bandwidth service. For instance, AT&T Wireless stopped selling this service in 2003 and discontinued it entirely in June 2004. CDPD existence was jeopardized by the huge and fast growing of digital cellular networks, where voice and data are treated in the same manner, and thus, can achieve higher rates at lower costs. In spite of its discontinuity, it is still worth studying CDPD, especially as concerns security. Although it was an advancement over AMPS, for example, we will see that the original CDPD project had several flaws in its security. A study of CDPD offers us a good opportunity to see how even commercially used systems may have security problems.

5.3.2 Basic Idea

CDPD was designed to operate during the idle time between AMPS calls. Although it used the same frequency as AMPS, CDPD itself was fully digital. It used a GMSK modulation to send data in the same frequency range as AMPS and a Reed-Solomon (63, 39) forward error correction to provide reliability against errors during the signal transmission through the RF channel.

CDPD exploits the long periods during which one or more of the radio channels within an AMPS cell sector are not used. CDPD uses these unused channels by hopping from one to the other. Thus, whenever a channel is required for voice traffic, the CDPD system chooses another unused channel and uses it to keep transmitting packets. If no channel is available for data traffic, CDPD will keep the digital data in a buffer and wait till an empty channel appears. Whenever the usage of a voice cellular network is typical, CDPD works fine. However, if it becomes overloaded less room is available for data traffic. Therefore, various CDPD providers have agreed to reserve some channels for CDPD in their networks.

5.3.3 Basic Infrastructure

As mentioned before, although CDPD uses the same frequency as AMPS, it requires separated hardware due to the fact it carries digital data. The basic components of the CDPD network are:

- End system (ES), including mobile end system (M-ES) and fixed end system (F-ES);
- Mobile data base station (MD-BS);
- Mobile data intermediate system (MD-IS);
- Intermediate system (IS).

These components will now be explained in detail. They are represented in Figure 5.16.

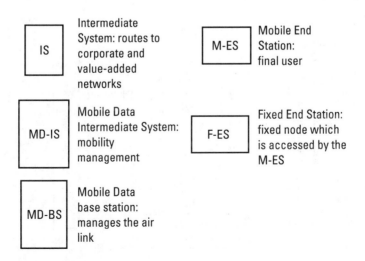

Figure 5.16 Components of the CDPD network.

There are also three basic interfaces in a CDPD network: the *A* interface, the *E* interface, and the *I* interface as shown in Figure 5.17.

The *A* interface is the RF link connecting the M-ES to the MD-BS. The *E* interface is a connecting point between a CDPD network and external networks. Conventional networking protocols are used here for establishing connections with external networks. The *I* interface is a connection point between different CDPD networks. As explained before, it is possible that a user of a CDPD managed by a certain provider moves to a location where the only CDPD service available is managed by a different provider. This is the need of *I* interfaces.

5.3.3.1 End Systems (ESs)

There are two kinds of end systems in a CDPD network: the mobile end system (M-ES) and the fixed end system (F-ES). End systems are the ultimate destination and source of data in a digital network. As its name implies, M-ES are nodes of the CDPD network that happen to be mobile. In the real world, they are used in police patrol cars, kiosks selling goods, or in commercial transportation trucks. They can be telemetry devices, personal communicators, or personal computers connected to a CDPD modem. The M-ES architecture consists of three functional blocks. These are the subscriber unit, the subscriber identity module, and the mobile application system, as shown in Figure 5.18.

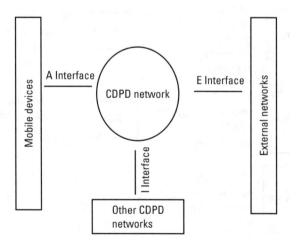

Figure 5.17 CDPD basic structure.

The subscriber unit is responsible for maintaining all communication between the M-ES and the rest of the CDPD network. It achieves the execution of all the air link protocols. The subscriber identity module possesses all the credentials and identification information necessary for authenticating the M-ES to the CDPD network. This function was separated from other modules of the M-ES in order to make it possible to use SIM cards, very much like in GSM systems. The mobile application subsystem is the part of the E-MS that manages its mobility and takes care of all the protocols from the network layer above.

The fixed end system (F-ES) is a typical node of a data network and can be either a node of a CDPD network (internal F-ES) or an external node (external F-ES). It can be, for example, the database of a city's criminals that is accessed by policemen in patrol cars.

5.3.3.2 The Mobile Data Base Station

The mobile data base station, together with the intermediate system, is responsible for connecting the M-ES with the rest of the CDPD network through an air link. It manages forward error correction, transmitter power parameters, and controlling access of several M-ES to a single radio channel. It also manages all the modulation data over the RF channel.

5.3.3.3 The Mobile Data Intermediate System

The mobile data intermediate system is responsible for all the mobility management in a CDPD network. It has two main functions: the

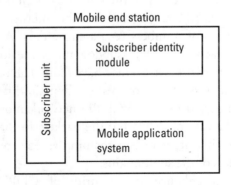

Figure 5.18 Mobile end station.

mobile serving function (MSF) and the mobile home function (MHF).

The mobile serving function provides end-to-end connection with a M-ES. When an M-ES announces itself to the system, it is registered by MSF. The mobile home function is responsible for receiving all the packets from the rest of a network directed to a certain mobile. It then forwards everything to the MSF, which then sends them to their final destination, the mobile unit.

5.3.3.4   The Intermediate System

The intermediate system (IS) is just a normal network router, responsible for connecting CDPD with other networks (external or internal). Typically, a packet traveling from a M-ES to a F-ES travels through several ISs.

5.3.4   How a CDPD Connection Works

Whenever a user buys an M-ES, he or she has to set up this device in order to be able to use the CDPD services. This setup process is very much like the registration of new handsets in cellular networks.

Each time an M-ES is turned on, it registers itself on the CDPD network. After the registration, an authentication and verification process begins, and, upon passing all these tests, the M-ES is connected to the CDPD network.

Each M-ES has a specific number that identifies it to the network. This number is known as the network entity identifier (NEI). The NEI is an address exactly in the same way as an Internet IP is an address. The NEI is 32 bits long and, like the Internet address, is represented by a series of numbers separated by dots. A subdomain is associated with each NEI. The subdomain is the location of the M-ES inside the network. M-ES may have more than one subdomain associated with it. If an M-ES moves to a domain different than the one originally assigned to it, the CDPD network mobility management should take care of delivering the packets to the correct location. This is possible if there is a roaming agreement between the different providers of the subdomains.

In order to enhance the security of CDPD systems, it was proposed that each mobile, after passing the verification and authentication tests, should agree on a random address, or, a nonce, which would identify the E-MS to the rest of the network (instead of the NEI). In this way, even if the nonce is intercepted by an eavesdropper, the original NEI would still be secure. This is called a temporary equipment identifier (TEI).

In the same way as the E-MS has its unique identification number, the CDPD carrier also is identified by a specific address. This address is called the service provider network identifier (SPNI) and it enables the mobile to identify in which domain it is located.

Once the M-ES is connected to a network, it can start to communicate. The following example shows how data is communicated from/to the M-ES. The main difficulty to overcome is the management of the mobility of the M-ES, which can move to a domain outside its home domain. Overcoming this difficulty is is very much like the roaming in normal cellular networks. CDPD encapsulates and forward Internet messages so that the M-ES appears to the rest of the network as it still resides behind the home domain.

Here we introduce two examples to clarify how the CDPD network manages mobility [9].

In our first example, a mobile user *A* wants to communicate with a user *B*. *A* is inside its home domain MD-IS(*A*). However, *B* is roaming in a domain different from its original MD-IS(B). Actually, *B* is inside *A*'s domain. When *B* sends a message to *A*, the following happens:

1. The packet sent by *B* goes from MD-ES(B) to the local MD-BS.
2. The packet goes from MD-BS to the controlling MD-IS(*A*), where the destination of the user is determined to be at home.
3. The packet goes from MD-IS(*A*) back to the local MD-BS.
4. The packets goes from MD-BS to M-ES(*A*), the final destination.

Figure 5.19 illustrates this process.

The mobility management in this case was relatively simple because the recipient of the message *A* was inside his home domain.

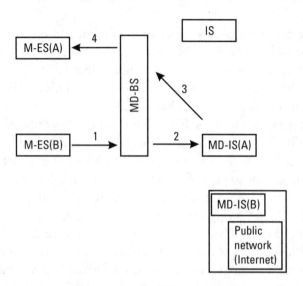

Figure 5.19 An M-ES outside its home domain contacting a local M-ES.

The situation becomes a little bit more complex when *A* sends a message to *B*, which is outside its home domain. The message goes through the following path:

1. The message is sent from M-ES(*A*) to local MD-BS.

2. The packet goes from MD-BS to the controlling MD-IS(*A*) where the location of the domain of the recipient of the message is determined.

3. MD-IS(*A*) routes through the gateway IS.

4 IS sends the packet to the Internet (and maybe other IS routers).

5 The packet arrives at the home domain of the recipient of the message MD-IS. MD-IS knows that user B is roaming on MD-IS(*A*) and routes the packet back there.

6. MD-IS(*A*) knows that the recipient of the message is inside its domain and sends the message to the correspondent MD-BS.

7. MD-BS sends the message to M-ES(B).

Figure 5.20 illustrates the process.

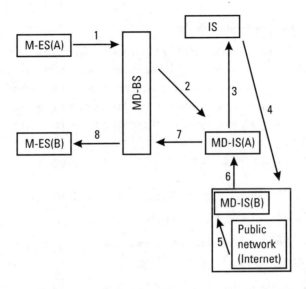

Figure 5.20 A local M-ES contacting an M-ES outside its home domain.

5.3.5 CDPD Security

The CDPD security was designed to achieve two goals: to protect the provider form fraudulent access and cloning of registered devices and to protect the user from casual eavesdropping.

The security protocols that were designed for the CDPD network achieve the following features:

- Data link confidentiality: All the information that is exchanged between the MD-BS and the M-SE (including the NEIs of M-ESs) is encrypted.

- M-ES authentication: Each NEI used by the M-ES is separately authenticated by the CDPD network to ensure that only the authorized possessor of the NEIs is using them.

- Key management: All secret keys required to operate the encryption algorithms are managed by the CDPD network.

- Access control: It is easy to restrict the access of a certain user to the CDPD network. These restrictions can be by location, screening lists, and so on.

5.3.5.1   The Authentication Protocol

To avoid cloning, the protocol challenges the M-ES each time it tries to register onto the network. The network asks the mobile two questions: (1) How many times have you been in this network? and (2) What was the password I gave you the last time you logged in? If the modem cannot answer both of these questions correctly, it is not allowed to register onto the network.

Before describing the protocol, first some notations:

- SHR_i : Shared historical record with time stamp i
- $SHR_i = (ARN, ASN)$, 80 bits
  - SHR_{i+1}: Shared historical record with time stamp i+1
- $SHR_{i+1} = (ARN, ASN+1)$, 80 bits
  - ARN: Authentication random number, 64 bits
  - ASN: Authentication sequence number, 16 bits
  - NEI: Network entity identification

1. The M-ES and the mobile serving function (MSF) of the MD-IS perform a Diffie-Hellman key exchange protocol (See Chapter 2) and agree upon a random key K_{MS}. Then M-ES and MSF use the key K_{MS} to create a secure channel between them. From now on, all the communication between M-ES and MSF is encrypted using K_{MS} RC4.
2. M-ES sends $Y_{MS} = E_{K_{MS}} (NEI, SHR_i)$ to MSF.
3. MSF decrypts $Y_{MS} = E_{K_{MS}} (NEI, SHR_i)$ and sends NEI, SHR_i to the mobile home function (MHF).

MHF checks to see if SHR_i and NEI are valid. If so, MHF sends (accept, SHR_{i+1}) or (refuse) back to MSF where MSF will accept or refuse the connection request made by M-ES.

The main advantage of this protocol is its simplicity. It is very easy to maintain and does not demand any additional infrastructure besides the CDPD network itself.

However, this protocol is vulnerable to the so-called man-in-the-middle attack. An adversary can pretend to be MSF and actively attack the CDPD network. As pointed out in [10], this can be accomplished by overpowering MSF by being closer to M-ES.

In addition, all the communication between MSF and MHF and, more generally, all the messages sent over the backbone network are not authenticated and not even encrypted. Thus an enemy obtaining access to these network links can obtain M-ES credentials.

5.3.5.2 A New Authentication Protocol

In [10] a new authentication protocol was proposed that fulfills the gaps of the previous protocol. It uses two authentication protocols embedded into one. One authentication takes place between M-ES and MHF and other between MSF and MHF. Following are some new notations before describing the protocol:

- $A_{xy}(I)$: Digital signature function, where X signs I and Y is able to verify this signature.

- S_{HM}: Common nonsecret state between MHF and M-ES. This state is usually the historical record from previous authentications.

- R_{HM}: Common nonce between MHF and M-ES.

- ID_S: Identification of MSF.

Four transmissions are performed in the protocol.

1. M-ES sends the following tuple to MSF:
   $$T_{MH} = (R_{MH}, T'_{MH})$$
   where $T'_{MH} = A_{MH}(S_{HM}, R_{HM}, ID_s = \text{``response''})$

2. Then MSF relays the following tuple to MHS:
   $$T_{SH} = (T_{MH}, R_{SH}, T'_{SH})$$
   where $T_{SH} = A_{MH}(T_{MH}, R_{SH}, ID_M = \text{``relay''})$

3. MHS then verifies the previous transmission. Upon acceptance, MHS sends the following tuple to M-ES routed through MSF:

$$T_{HS} = (S'_{HM}, T'_{HM}, T'_{HS})$$

where $T'_{HM} = A_{HM}(R_{MH}, S'_{HM}.ID_S = \text{"refresh"})$

and $T'_{HS} = A_{HS}(R_{SH}, T'_{HM}.ID_M = \text{"accept"})$

4. Finally, MSF relays the following message to M-ES

$$T_{HM} = (S'_{HM}, T'_{HM}, \text{"refresh"})$$

where $T'_{HM} = A_{HM}(R_{MH}, S'_{HM}, ID_H = \text{"refresh"})$

This protocol is robust against man-in-the-middle attacks, because mutual trust is established between M-ES, MSF and MHF. Moreover, it is a trivial task to modify this protocol in order to obtain a key agreement protocol and to make it anonymous. However, there are a few drawbacks.

- The key management problem becomes a serious issue in this new protocol.

- MHF should keep secret keys for each M-ES and MSF in the network.

- Likewise, each MSF must maintain keys for each MHF.

- This protocol would never scale up in a large network, such as the Internet.

- There is no practical way to update the keys (in case they are compromised) for the mobiles.

- Possible intrusion attempts by M-Es are not detected until the step 2 of the protocol, thus backbone network bandwidth is wasted. Moreover, the system could be susceptible to denial of service (DoS) attacks.

Thus, the last points show us that, when designing cryptographic protocols for real-world applications, one has to carefully balance the trade-off between security and performance of the cryptosystem.

References

[1] Sandberg, J., "Flaw Is Found in Digital Phone System that May Let Hackers Get Free Service," *The Wall Street Journal*, April 13, 1998.

[2] Golic, J. D., "Cryptanalysis of Alleged A5 Stream Cipher," http://jya.com/a5-hack.htm.

[3] Biryukov, A., Shmair, A., and Wagner, D., "Real Time Cryptanalysis of A5/1 on a PC," *Lecture Notes on Computer Science, LNCS1978*, B. Schneier, (ed.), Springer, 2000, pp. 1–18.

[4] Rao, J. R., et al., "Partition Attacks: Or How to Rapidly Clone Some GSM Cards," *Proc. 2002 Symposium on Security and Privacy (S&P2002)*, Berkeley, CA, May 12–15, 2002, pp. 31–44.

[5] Freier, A. O., Karlton, P. and Kocher, P. C., "The SSL Protocol Version 3.0," draft-freier-ssl-version3-02.txt, accessed November 1996.

[6] http://www.netscape.com/eng/security/ssl_2.html.

[7] Housley, R., et al., "Internet X.509 Public Key Infrastructure Certificate and Certificate Revocation List (CRL) Profile," RFC3280, April 2002.

[8] Luotonen, A., "Tunneling SSL Through a WWW Proxy," draft-luotonen-ssl-tunneling-02.txt, December 1995.

[9] http://www.sierrawireless.com/pub/doc/2130006.pdf.

[10] Frankel, Y., et al., "Security Issues in a CDPD Wireless Network," *IEEE Personal Communications*, Vol. 2, No. 4, August 1995, pp. 16–27.

6

Security in 3G and 4G Systems

In the previous chapter, security considerations and mechanisms of second generation mobile systems were discussed. Here, we discuss the security of third generation systems (3G) and briefly introduce fourth genereation wireless communication systems (4G). 3G systems bring to the user multimedia communications, mobile commerce, among many other applications, in the wireless environment.

6.1   3G Wireless Communications Systems

Third generation mobile communication systems are called by several different acronyms 3G, UMTS, IMT2000 and W-CDMA, to name a few. The promises of third generation mobile phones are fast Internet surfing, advanced value- added services and video telephony. 3G includes capabilities such as enhanced multimedia (voice, data, and video), a wide range of services (i. e. e-mail, paging, fax, video-conferencing, and web browsing), broad bandwidth, high data transmission speed (up to 2 Mbps), routing flexibility (repeater, satellite, LAN), operation at 2 GHz transmit and receive frequencies, and roaming capability throughout Europe, Asia and North America.

3G technology improves upon 2G systems in many ways. For example, it moves towards packet switching from circuit switching. Packet switching uses the communication system more effectively, therefore boosting its capacity. Packet switching also enables users to be online continuously. Via careful use of the frequency spectrum and inventive coding methods, 3G technology is supposed to achieve bit rates up to 2 Mbps. Essential qualities and characteristics of a 3G wireless system are:

- Bit rates reaching up to 2 Mbps;
- Variable bit rate to offer bandwidth on demand;
- Multiplexing of services with different quality requirements on a single connection, (e.g. speech, video, and packet data);
- Delay requirements from delay-sensitive real-time traffic to flexible best-effort packet data;
- Quality requirements from 10% frame error rate to 10^{-6} bit error rate;
- Coexistence of second and third generation systems and inter-system handovers for coverage enhancements and load balancing;
- Support of asymmetric uplink and downlink traffic (e.g., Web browsing causes more loading to downlink than to the uplink);
- High spectrum efficiency;
- Coexistence of FDD and TDD modes.

Key considerations for 3G wireless communications are:

- High-speed access, supporting broadband services such as fast Internet access or multimedia-type applications;
- Competitive prices (expected to be eventually at same price range or cheaper than current cellular systems);
- Compatibility with existing cellular infrastructure, thus offering an effective evolutionary path for existing wireless networks.

3G Systems are intended to provide a global mobility with wide range of services including telephony, paging, messaging, Internet and broadband data. International Telecommunication Union (ITU) started the process of defining the standard for third generation systems, referred to as International Mobile Telecommunications 2000 (IMT-2000). In Europe, European Telecommunications Standards Institute (ETSI) was responsible of UMTS standardization process. In 1998, Third Generation Partnership Project (3GPP) was formed to continue the technical specification work.

6.2 Third Generation Partnership Project (3GPP)

The Third Generation Partnership Project is developing technical specifications for third-generation standards. 3GPP is a global cooperation between six organizational partners (ARIB, CCSA, ETSI, T1, TTA and TTC) who are recognized as being the world's major standardization bodies from Japan, China, Europe, the United States, and Korea.

The original scope of 3GPP was to produce globally applicable technical specifications and technical reports for a third generation mobile system based on evolved GSM core networks and the radio access technologies that they support (i.e., Universal Terrestrial Radio Access (UTRA) both Frequency Division Duplex (FDD) and Time Division Duplex (TDD) modes). The scope was subsequently amended to include the maintenance and development of the Global System for Mobile communication (GSM) Technical Specifications and Technical Reports including evolved radio access technologies (e.g. General Packet Radio Service (GPRS) and Enhanced Data rates for GSM Evolution (EDGE)).

6.2.1 3GPP Security Objectives

3GPP security was based on GSM security. The whole 3G security was designed based on three fundamental principles:

1. The security for 3G will build on the security features of 2G systems. Some of the robust features of 2G systems will be retained.

2. The 3G security will improve on the security of the 2G systems. Some security holes and disadvantages of 2G systems will be addressed and corrected in 3G systems.

3. 3G security will offer new features and will secure new services offered by 3G.

The general objectives for 3G security features are [1]:

- To ensure that information generated by or relating to a user is adequately protected against misuse or misappropriation.

- To ensure that the resources and services providing by serving networks and home network environments are adequately protected against misuse and misappropriation.

- To ensure that the security features standardized are compatible with worldwide availability.

- To ensure that the security features are adequately standardized to ensure worldwide interoperability and roaming between different serving network.

- To ensure that the level of protection afforded to users and providers of services is better than that provided in contemporary fixed and mobile networks (including GSM).

- To ensure that the implementation of 3G security features and mechanisms can be extended and enhanced as required by new threats and services.

6.3   3G Security Architecture

It is necessary to look into the security issues within the security architecture from various levels. Figure 6.1 provides an overview of the complete 3G security architecture [2].

The 3G specifications for security define five different security features.

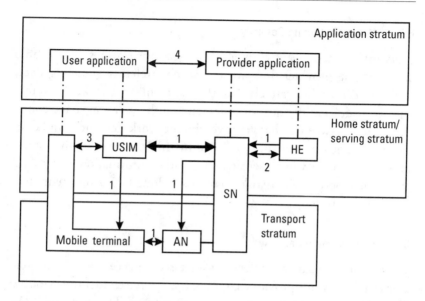

Figure 6.1 3G security architecture.

6.3.1 Network Access Security

The set of security features that provide users with secure access to 3G services and which in particular protect against attacks on the (radio) access link. This feature provides user identify confidentiality, authentication of users, confidentiality of data on the network access link, date integrity, and mobile equipment identification. User identity confidentiality is achieved by the use of temporary identities. The transmission of the International Mobile User Identity (IMUI) over the air interface in clear text is avoided as far as possible. Authentication of users is achieved by means of mutual authentication between the user and the network using secret key through the challenge-response mechanism. A secret Cipher Key (CK) is established as part of the Authentication and Key Agreement (AKA) process to provide data confidentiality. Data integrity is a new security feature included in 3G systems. The 3G Integrity Algorithm along with an Integrity Key (IK) will be used for providing data integrity. The IK is established as a part of the AKA process. Mobile equipment identification is done by using an International Mobile Equipment Identifier (IMEI) that uniquely identifies mobile equipment.

6.3.2 Network Domain Security

Network domain security is a set of security features that enable nodes in the provider domain to securely exchange signaling data, and protect against attacks on the wire line network. This feature provides entity (network element) authentication, data confidentiality (between exchanges involving network elements), and data integrity. The functionality provided by this feature is highly important in the case where sensitive signaling information has to be exchanged between network elements belonging to different network elements.

6.3.3 User Domain Security

User domain security are features that secure access to mobile stations. This feature provides User to User Services Identity Module (USIM) and USIM to terminal authentication. The user to USIM authentication is accomplished by the means of a secret that is stored securely in the USIM. The user can have access to the USIM only if he/she proves knowledge of the secret.

The USIM and the terminal share a secret that is stored securely in USIM and the terminal. To gain access to the terminal, the USIM has to prove knowledge of the secret.

6.3.4 Provider-User Link Security

This set of security features enable applications in both the user and in the provider domain to securely exchange messages. The 3G systems will provide the capability for operators or third party providers to create applications, which are resident on the USIM. Hence there exists a need to secure messages, which are transferred over the network to applications on the USIM with the level of security chosen by the network operator or the application provider. The features provided to ensure security of messages are:

- Entity Authentication of Applications;
- Data Origin Authentication of Application Data;
- Data integrity of Application Data;
- Replay Detection of Application Data;

- Sequence Integrity of Application Data;
- Proof of Receipt.

6.3.5 Visibility and Configurability of Security

This set of security features informs users as to whether or not a security feature is in operation and whether the use and provision of services should depend on the security feature. Ideally, all the security features should be transparent to the user. In some cases, and in accordance with the user's concern, the user must be provided greater visibility into the operation of the security features, as for example, indication of the access network encryption, indication of the level of security.

Though the security levels provided by 3G systems exceed by far those provided by the earlier cellular telecommunication systems, the robustness of 3G systems with respect to the security features are yet to be tested as 3G systems are deployed in the market very recently.

6.4 Authentication and Key Agreement (AKA) in 3GPP

The design of 3GPP security is such that not every carrier has to use the same algorithms for authentication and key exchange (AKA). Therefore, instead of producing a standard, 3GPP publishes a recommendation or example algorithm for those providers who do not wish to design and implement their own algorithms. The current suggestion for AKA algorithms is called MILENAGE. MILENAGE is a construction based on AES, or Rijndael. MILENAGE was chosen for its believed strength, its ability to run very fast on the chipsets present in cell phones, and its low memory usage. (For a description of AES, please see Chapter 2.)

3GPP AKA provides mutual authentication for both the user terminal (MS) and the network, showing the knowledge of a secret key which is shared between and available only to the USIM and the AuC in the user's home location register (HLR). During AKA the user and the network authenticate each other and also they agree on the cipher key (CK) and integrity key (IK). CK and IK are used until their time expires. The data flow for AKA protocol is depicted in the Figure 6.2.

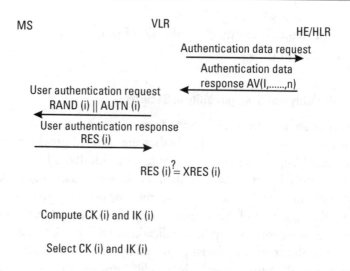

Figure 6.2 Authentication and key agreement data flow.

Upon receipt of the authentication data request from the VLR, the HE/AuC sends an array of n authentication vectors to the VLR. The authentication vectors are ordered according to a sequence number. Each vector consists of the following components: a random number (RAND), an expected response (XRES), a cipher key (CK), an integrity key (IK), and an authentication token (AUTN). Each authentication vector is used in only one authentication between VLR and USIM.

During the AKA process VLR selects the next authentication vector from the ordered array and sends the parameters RAND and AUTN to the user. If the received AUTN in USIM is the valid token then it produces a response RES and sends back to VLR. VLR compares the received RES with XRES and if they match, VLR considers AKA to be successfully completed. In VLR, CK and IK stored in the corresponding authentication vector are used for ciphering and integrity functions. The USIM also computes CK and IK, which are used for ciphering and integrity functions.

6.5 Confidentiality and Integrity

One of the goals of 3G system is to provide global roaming. Hence, the air interface encryption and integrity algorithms must be the same for all carriers (allowing a subscriber to roam into another carrier's territory and still be able to communicate with the base stations located there). The 3GPP algorithms used for these purposes are both built around a block cipher called KASUMI. KASUMI was designed by the Security Algorithms Group of Experts (SAGE), part of the European standards body ETSI. Rather than invent a cipher from scratch, an existing algorithm, MISTY1, was selected by SAGE and slightly optimized for implementation in hardware. KASUMI is the Japanese word for "misty."

KASUMI is a block ciphering algorithm that uses a 128 bit key and generates a 64 bit output stream from a 64 bit input stream. KASUMI is essentially a Feistel Cipher with eight rounds. The 3GPP performed extensive analysis on KASUMI, including linear cryptanalysis, differential cryptanalysis, higher order differential cryptanalysis, and weak key searches. The 3GPP report on KASUMI states that no weak keys were discovered and no practical attacks could be found that can break KASUMI at its full eight round implementation.

6.5.1 Confidentiality

The $f8$ algorithm is used to protect the user and signaling data sent over the radio access link between radio network controller (RNC) and mobile station (MS). The $f8$ algorithm is based on the Kasumi algorithm. $f8$ is used to encrypt plaintext by applying a keystream using a bitwise XOR operation. The input parameter to the algorithm are the cipher key (CK), a time dependent input (COUNT-C), the bearer identity (BEARER), the direction of transmission (DIR) and the length of the keystream required (LEN). Based on the input parameters the algorithm generates the output keystream block, which is used to encrypt the plaintext block to produce the ciphertext.

The plain text can be recovered by generating the keystream using the same input parameters and applying it to the received cipertext using bitwise XOR operation. Figure 6.3 depicts both encryption and decryption mechanism for providing user and signaling data confidentiality.

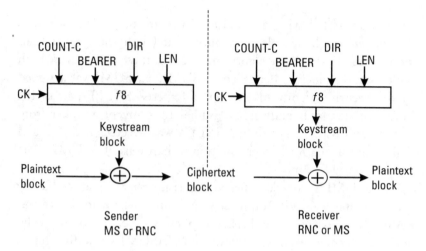

Figure 6.3 Air interface confidentiality mechanism.

6.5.2 Data Integrity

False base station attack is the major problem in the second generation GSM system. To protect against false base station attacks, the
receiving entity, (MS or SN, must be able to verify that the signaling
data has not been modified in an unauthorized way since being sent
by the sending entity (SN or MS). It must also be ensured that the
data origin of the signaling data received is indeed the one that
itclaimed to be. This is achieved by inclusion of data integrity function for the signaling data. The message authentication code (MAC)
function $f9$ is used to authenticate data integrity and data origin of
signaling data transmitted between the MS and the radio network
controller (RNC). The MAC function $f9$ is allocated to the MS and
RNC. The input parameters to the algorithm are the integrity key
(IK), a time dependent input (COUNT-I), a random value generated
by the network side (FRESH), the direction of transmission (DIR)
and the signaling data (MESSAGE). Based on these input parameters
the sender (MS or RNC) computes the message authentication code
for data integrity (MAC-I) using $f9$ algorithm. MAC-I is then
appended to the message when sent over the radio access link. The
receiver computes XMAC-I on the message received using the same
parameters as the sender using $f9$ algorithm. Figure 6.4 depicts the

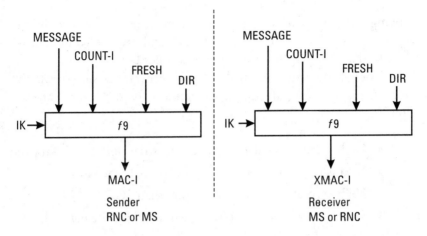

Figure 6.4 Signaling data integrity mechanism.

derivation of MAC-I and XMAC-I for air interface integrity mechanism.

6.6 4G Wireless Communications Systems

Cellular service providers are slowly beginning to deploy third generation (3G) cellular services. As access technology increases, voice, video, multimedia, and broadband data services are becoming integrated into the same network. However, 3G as a true broadband service, has not quite lived up to its expectations. It is not fully integrated with the Internet and an increasing demand for extremely high-bandwidth services made 3G systems somehow obsolete even before they became fully operational. While 3G hasn't quite arrived, designers are already thinking about a 4G technology. To achieve the goals of true broadband cellular service, the systems have to make the leap to a fourth generation (4G) network. 4G is intended to provide high speed, high capacity, low cost per bit, IP based services. The goal is to have data rates up to 20 Mbps. Most probably, the 4G network would be a network which is a combination of different technologies (current cellular networks, 3G cellular network, wireless LAN, and so forth) working together using suitable interoperability protocols, as for example, Mobile IP. Standardization work on 4G has already begun. The network is expected to support communications to moving vehicle up to speeds of 250 km/hr. The aim is to bring together 4G mobile technology, WLAN, and satellite communications so that they can all work together seamlessly.

Though the platform for the fourth generation is not yet been considered, following are the considerations for the future generation [3]:

- High speed (vehicular: 2 Mbps, pedestrian/indoor: 20 Mbps);
- Next generation Internet support (IPv6, QoS, Mo-IP);
- High-capacity 5 to 10 times higher than 3G;
- Seamless services with fixed network and private network;
- Flexible for providing new services;
- Utilize higher frequencies (microwave: 3 to 8 GHz);
- Lower system cost (one-tenth of IMT2000).

In addition to these, security is also a prime consideration.

References

[1]   3G TS 33.120, "3G Security; Security Principles and Objectives."

[2]   3GPP TS 33.102, "3G Security; Security Architecture."

[3]   Nakajima, N., "Future Mobile Communications Systems in Japan," *Wireless Personal Communications*, Vol. 17, No. 2–3, 2001, pp. 209–223.

7

Wireless Application Protocol (WAP)

7.1 Introduction

The Wireless Application Protocol (WAP) is an open specification that enables mobile users to have access to the Internet. WAP specifies both communication protocols and application environment so that it can work regardless of the underlying wireless networks, such as CDPD, CDMA, GSM, PDC, PHS, DECT, and GPRS, and can be built over any operating system including PalmOS, Windows CE, JavaOS and so on.

The first generation of WAP is referred to as WAP 1.x (or WAP1). The initial version WAP 1.0 was released in 1998. The next generation is WAP 2.x (or WAP2). WAP 2.0 was released in January 2002. The main difference between WAP1 and WAP2 is summarized as follows: WAP2 assumes relatively high-performance mobile terminals and employs a lot of Internet standards. This enables WAP2 mobile terminals to interact with servers in the Internet directly and then to establish secure channels with them end-to-end. On the other hand, WAP1 employs optimized protocols for relatively inexpensive terminals and low-bandwidth wireless networks while sharing part of the tasks with WAP gateways. This enables mobile terminals to be simple, but secure connections must be severed by the WAP gateways to exchange WAP1 protocols with the Internet protocols. The details are described in this chapter.

The organization of this chapter is given as follows: In Section 7.2, the protocol stacks of WAP1 and 2 are explained. In Section 7.3, PKI (Public-Key Infrastructure) model for WAP is given. WTLS (Wireless TLS) and WAP profiled TLS are explained in Section 7.4 and 7.5, respectively. Then WIM (WAP Identity Module) is explained in Section 7.6. Finally, current status and further information on WAP are summarized in Section 7.7.

7.2 WAP Protocol Stack

This chapter describes the protocol stacks of WAP1 and 2. Two typical protocol stacks of WAP1 and 2 are illustrated in Figures 7.1 and 7.2, respectively.

In both figures, each layer provides the following functionalities:

- *Wireless Application Environment (WAE):* WAE provides an environment for processing WAP applications, usually Web applications.

- *Wireless Session Protocol (WSP):* WSP provides similar functionality to HTTP/1.1 [1,2] with new features, such as long-lived sessions, session suspend/resume and transformation of application data. Data transformation is done so that the size can be smaller and that the WAP1 clients can process

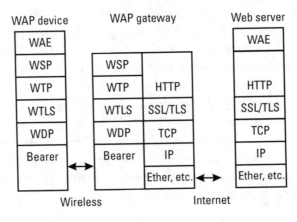

Figure 7.1 Typical protocol stack of WAP1.

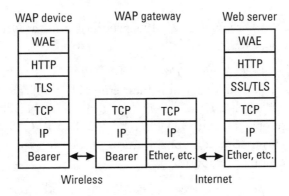

Figure 7.2 Typical protocol stack of WAP2.

it with smaller complexity. Usually, the WAP gateway transforms html (or wml) to the corresponding binary file, which
is smaller and easier to process for WAP1 clients. WSP has
a consistent interface for two session services: connec-
tion-mode services over the transaction layer protocol and
connectionless services over a secure or nonsecure datagram
transport.

- *Wireless Transaction Protocol (WTP):* WTP provides reliability of transmission data. It is designed for low-performance
mobile clients so that it can manage acknowledgments and
retransmission of lost data efficiently over the wireless
datagram network.

- *Wireless Transport Layer Security (WTLS):* WTLS provides
confidentiality, integrity and authentication between a WAP
1.x client and a WAP Gateway. It also protects against replay
attacks following similar trends from TLS [27]. Additional
features of WTLS to TLS include datagram support, optimized handshake, dynamic key refreshing, elliptic curve
cryptosystem support and so on.

- *Wireless Datagram Protocol (WDP):* WDP provides a general
datagram service, which offers a consistent service to the
upper layer protocols and communicates transparently over
one of the available underlying bearer services. This consis-
tency is provided by a set of adaptations to specific features
of these bearers. Usually, UDP/IP or SMS (Short Message

Service) are used as WDP depending on the available under-lying bearer services.

- *Hypertext Transfer Protocol (HTTP):* HTTP is a protocol to request and transmit files, especially for Web pages and Web applications. A wireless profile of HTTP is fully interoperable with HTTP/1.1 [1, 2] and RFC2818 [3] and then supports additional features, such as message body compression of responses.

- *Transport Layer Security (TLS):* TLS provides *confidentiality,* data integrity and authentication between two entities. In WAP2, the connection is end to end, i.e. the WAP 2.0 client has a direct connection with the application server, and hence the WAP Gateway proxies nothing but encrypted TLS data. In WAP1, TLS is used between the WAP Gateway and the application server. The TLS specification within WAP has been profiled from RFC 2246 optimizing cipher suites, cer-tificate formats (that are carried out within a separate specifi-cation certProf [4]), and other issues such as session resume.

- *Transmission Control Protocol (TCP):* TCP provides reliability of transmission data over the IP layer. It manages sequence and acknowledgement numbers and re-transmits lost data. A wireless profile of TCP defines optimized parameters for the wireless environment maintaining full interoperability with standard TCP implementations in the Internet.

- *Internet Protocol (IP):* IP is a protocol to manage routing of transmission data, addressing of hosts/networks and so on.

As shown in Figure 7.1, application data in WAP1 are exchanged by way of WAP Gateways that recognize both Internet and WAP protocols. WAP Gateways take on a part of jobs for WAP1 devices so that the WAP1 devices can be simpler and the communica-tion data over the air can be reduced. The drawback is that the secure connection is cut on the WAP Gateways for transforming application data. On the contrary, WAP2 devices recognize Internet standard protocols (or their wireless profiles), and then establish TLS secure connections to servers end-to-end as shown in Figure 7.2. The main

Table 7.1
Certificate Support versus WTLS Class

Classes/Entities	WAP Devices	Servers and WAP Gateways
WTLS Class1	Optional	Optional
WTLS Class2	Optional	Mandatory
WTLS Class 3 (and SignText)	Mandatory	Mandatory

functionality of a WAP proxy is to adapt the profile of TCP for the air independently of the upper layer, TLS.

7.3 WAP PKI Model

This section describes public-key infrastructure (PKI) models in WAP. PKI is an infrastructure to authenticate public-keys, that is, to identify the owner of the secret-key corresponding to the public-key. This is usually done by a trusted third party, certificate authority (CA), issuing digital certificates and then by users verifying the issued digital certificates. A digital certificate contains a public key, a distinguished name of the owner of the public key and so on. And then they are digitally signed by a CA. The ownership verification of the public keys is crucial over the insecure networks, such as Internet, since anyone can create a key claiming that this is the key of someone else.

PKI models in WAP can be divided into the three classes according to which entities have their certificates as shown in Table 7.1. In WTLS Class1, entities do not necessarily have their certificates or support certificate processing (PKI in general). Thus the ownership of the keys is not guaranteed by the infrastructure, which means this class is vulnerable to the man-in-the-middle attack unless the ownership of the keys is guaranteed with some methods, such as by verifying digests of the exchanged keys or by using a preshared secret. This class, however, still provides confidentiality against eavesdropping, thus it can be used in circumstances where use cases permit communication with unauthenticated sources but data confidentiality and integrity being a requirement against eavesdroppers. In

WTLS Class 2, servers and WAP gateways have their certificates, and can be identified by anyone using the infrastructure. In WTLS Class 3 (and SignText), not only servers and WAP Gateways but also WAP devices have their certificates, and therefore, can mutually authenticate each other. Bellow is a precise description of WTLS Classes 2 and 3. (WTLS Class 1 is omitted since it is obtained by removing certificates from WTLS Class 2 and 3.)

7.3.1 WTLS Class 2

WTLS Class 2 is a PKI model that mandates server side authentication, requiring servers to acquire server certificates to authenticate themselves to the WAP client. This class is useful if no client authentication is required, clients are not capable of authenticating themselves to the server (i.e. client not supporting private key operations), or if client authentication has already been equipped. The former example is that a server publishes information to the public over the Internet and WAP devices ensure an authenticated server is verifying the integrity of the data. Since the data are public no client authentication is required even though both the server and the downloading data should be authentic. In WAP1, this server authentication is done in two phases, the server authentication by the WAP gateway and the WAP gateway authentication by the WAP device. In this instance, an application, or a mail server, equips a password authentication mechanism that simply receives a password and then checks it. Since the client authentication is provided by the application, PKI does not need to provide it while a secure channel to the server must be provided for protecting passwords against eavesdroppers.

WTLS Class 2 can then be divided into two types according to whether WAP1 or 2 is employed. In WAP2, a WAP device directly authenticates a server and then establishes secure channels to the authenticated server using a WAP profile of TLS. This end-to-end secure connection is referred to as "end-to-end security." In WAP1, a WAP device firstly authenticates a WAP gateway and then establishes secure channels to the gateway using WTLS. Then the WAP gateway authenticates the server, to which the WAP device requests a connection, and finally, the WAP gateway establishes secure channels between the gateway and the server using SSL/TLS. This is referred to as "two-phase security," in contrast with the end-to-end security in WAP2. Two-phase security is less secure than end-to-end security since the communication in it can be eavesdropped or altered on the gateway. Precise steps and diagrams of WTLS Class 2 are given in Figures 7.3, 7.4, 7.5, and 7.6, respectively.

Setup for Servers and WAP Gateways

First of all, administrators of servers and WAP Gateways must obtain their PKI certificates so that they can be authenticated over the network. A typical way to do this is described as follows:

(Ss1) An administrator generates a pair of a public key and a private key, and a request form to obtain a certificate. The certificate-request includes the information on the public key itself, a distinguished server (or gateway) name corresponding to the public key and so on. The administrator sends it to a PKI portal with some documents proving the identity of the server (or the gateway). A PKI portal may be a RA and/or a CA recognizing WAP profile of PKI.

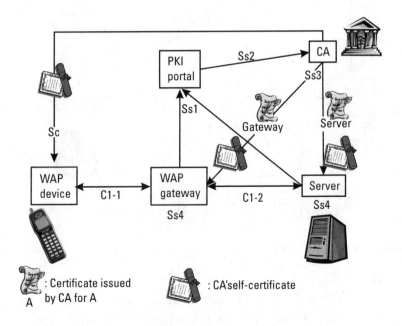

Figure 7.3 WAP1 in WTLS Class 2.

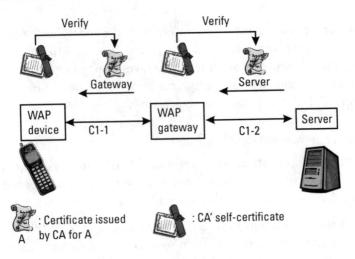

Figure 7.4 WAP1 secure connection in WTLS Class 2.

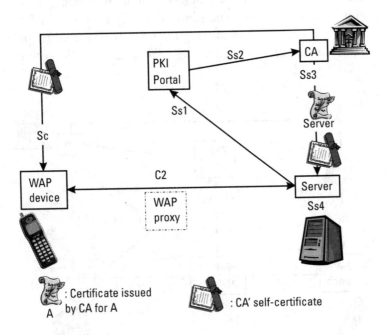

Figure 7.5 WAP2 in WTLS Class 2.

(Ss2) The PKI portal confirms the identity of the server (or the gateway) in the certificate request and forwards the request to the CA.

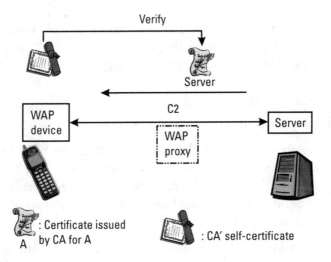

Figure 7.6 WAP2 secure connection in WTLS Class 2.

(Ss3) The CA issues a PKI certificate and sends it to the re-quester. A certificate includes the information on the public key, the distinguished name of the server (or the gateway), the name of hash and signature algorithms and so on, and then they are signed by the CA.

(Ss4) The administrator obtains both the issued certificate and the CA's self-certificate (or a certificate chain from a root CA to the issued certificate). The CA's self-certificate includes the public key of the CA, which is used to verify the CA's digital signatures. Self-certificates must be obtained in an authentic way since there is no way to verify them. This means a way where no modification or no replacement is feasible, either by hand or by a certified snail mail. Finally the administrator verifies the signature in the issued certificate using the CA's public key and then installs both the issued certificate and the CA's self-certificate to the server (or the gateway).

Setup for WAP Devices

The following describes WAP device setup:

(Sc) Each WAP device obtains self-certificates of some root CAs (or certificate chains from the root CAs) in an authentic way so that it can verify the certificates issued by the CAs later on. Users may be unaware of this acquisition process, since it be preinstalled or installed later with an update.

Secure Connection in WAP1

The following describes secure connections in WAP1:

(C1-1) Assuming that the WAP device has the WAP Gateway root certificate and end entity certificate installed, the WAP device verifies the certificate of the WAP Gateway using a CA's certificate, and then establishes secure WTLS channels to the authorized gateway.

(C1-2) The WAP Gateway verifies the certificate of the server, to which the WAP device requests to connect, using a CA's certificate, and then establishes secure SSL/TLS channels to the authorized sever.

Secure Connection in WAP2

The following descibes secure connection in WAP2:

(C2) A WAP device verifies the certificate of the server using a CA's certificate, and then verifies that the signature in the TLS handshake is indeed signed by the content server and then establishes secure TLS channels to the authorized sever. The connection may be established by way of WAP Proxy, but it is simply to change the profile of communication in the lower levels of the protocol stack than TLS.

7.3.2   WTLS Class 3 and SignText

Both WTLS Class 3 and SignText provide WAP devices with a functionality of issuing digital signatures that can be used for client authentication and non-repudiation of contracts. The difference between WTLS Class 3 and SignText is described as follows: WTLS Class 3 is used in the transport layer and provides additional functionalities to WTLS Class 2 whereas SignText is a standalone mechanism at the application layer, which can be invoked by WMLScript and so on.

This class requires a WAP device to have its public-key private-key pair and its certificate issued by a CA in addition to the procedures in WTLS Class2. This set-up phase may be performed at a purchase of the WAP device or prompted later on by a server requesting a tight authentication, for example, for a banking application. There are two ways to do this according to whether or not the device is memory constraint. Precise steps and the diagrams are given in Figures 7.7, 7.8, 7.9, and 7.10, respectively.

Setup for Memory Nonconstraint WAP Devices

The following descibes setup for memory nonconstraint WAP devices:

(Ss1) A WAP device generates a pair of a public key and a private key. (Or the pair is generated outside of the device and later installed). The device creates a certificate-request and then sends it to a PKI portal.

(Ss2) The PKI portal confirms the identity of the user in the certificate request and forwards the request to a CA.

(Ss3) The CA issues a PKI certificate and sends it to the requester.

(Ss4) The WAP device obtains the issued certificate, verifies the signature in it using the CA's public key, and then installs it to the WAP device.

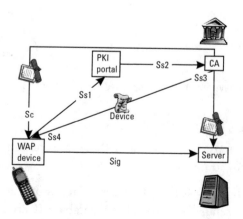

Figure 7.7 WTLS Class3 and SignText for a memory nonconstraint device.

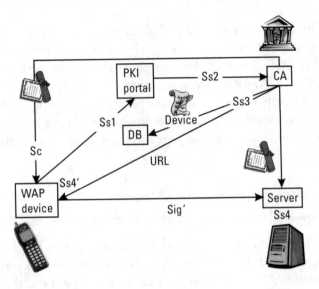

Figure 7.8 WTLS Class3 and SignText for a memory constraint device.

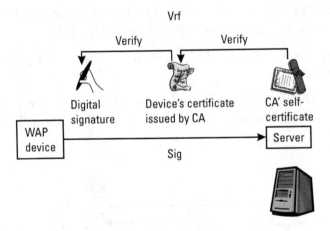

Figure 7.9 Sign and verification for a memory nonconstraint device.

Setup for Memory Constraint WAP Devices

The first two steps, Step (Ss1) and (Ss2), are the same as those for memory nonconstraint devices:

> (Ss3') The CA issues a PKI certificate, stores it in a database and then sends the URL of the database (Certificate URL) to the device.

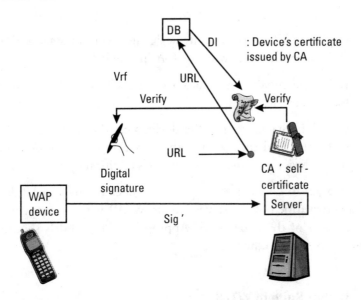

Figure 7.10 Sign and verification for a memory constraint device.

(Ss4') The WAP device obtains the CertificateURL instead of the issued certificate, and then stores it in his device. This saves both the necessary memory on the WAP device and the transmission data. Of course, the device may verify the issued certificate after obtaining it from the CertificateURL.

Sign and Verification for Memory Nonconstraint Devices

The following descibes sign and verification for memory nonconstraint devices:

(Sig) A WAP device generates a signature on a message and then sends it to the server with the device's certificate.

(Vrf) The server verifies the device's certificate using a CA's certificate, and then verifies the signature using the public-key of the device included in the device's certificate.

Sign and Verification for Memory Constraint Devices

The following describes sign and verification for memory constraint devices:

(Sig) A WAP device generates a signature on a message and then sends it to the server with the CertificateURL.

(Dl) The server downloads the device's certificate from the CertificateURL.

(Vrf) The server verifies the device's certificate using the CA's certificate, and then verifies the signature using the public-key of the device included in the device's certificate.

A point to notice is that signatures generated by a WAP device are directly verified by the server even in WAP1 where secure connection consists of the two phases. Appropriate PKI models should be chosen according to the needs.

7.4 Cipher Suite in WTLS

As mentioned in Sections 7.2 and 7.3, WTLS is a protocol layer that provides secure functionalities. The available functionalities and the security level, however, depend on the cipher suite negotiated at the beginning of the WTLS or TLS handshake. A cipher suite is a suite of crypto primitives, such as a key-establishment protocol, a bulk encryption algorithm and a MAC algorithm. While a cipher suite in SSL/TLS defines a combination of them, they can be chosen independently in WTLS. This allows flexible combination of the crypto primitives. Properties of each crypto primitive are summarized next.

7.4.1 Key Exchange Suite

A key exchange suite defines a key-establishment protocol, which is a protocol for sharing a fresh secret between a client and a server (or among group members). The available key exchange suites in WTLS are summarized in Table 7.2.

In Table 7.2, NULL means no key-exchange is performed. Since this provides no security, one should avoid choosing it. SHARED_ SECRET is a simplified handshake where both a client (a WAP device) and a server share a secret key in advance and then use it as a premaster key in the initial stages of the WTLS/TLS handshake. The size of the secret key should be long enough, (no less than 80

Table 7.2

Key Exchange Suites in WTLS

Key Exchange Suite	Assigned Number
NULL	0
SHARED_SECRET	1
DH_anon	2
DH_anon_512	3
DH_anon_768	4
RSA_anon	5
RSA_anon_512	6
RSA_anon_768	7
RSA	8
RSA_512	9
RSA_768	10
ECDH_anon	11
ECDH_anon_113	12
ECDH_anon_131	13
ECDH_ECDSA	14
ECDH_anon_uncomp	15
ECDH_anon_uncomp _113	16
ECDH_anon_uncomp _131	17
ECDH_ECDSA_uncomp	18

bits), and must not be a short password that can be cracked with exhaustive search. This suite is useful if a client and a server can share a long secret in advance off-line, either by hand or by a snail mail.

*_anon, such as DH_anon, RSA_anon, and ECDH_anon, is the key exchange suite without authentication. In these cases, raw keys are derived and used to transfer bulk encryption secrets. These suites should not be used since they are vulnerable to the man-in-the-middle attack where an adversary simultaneously communicates with both a server and a client impersonating the other entity to each. The man-in-the-middle attack is possible if authentication is not provided. It is not very difficult over the Internet to guide a client to a fake server by controlling DHCP (Dynamic Host Configuration Protocol) and/or DNS (Domain Name System). Thus, *_anon should not be used unless the exchanged public-key is verified with a certain method, such as by comparing the digest with the unmodified version of it given in an authentic way, such as by hand. In addition, *_anon does not provide perfect anonymity while "anon" is the abbreviation of "anonymous." If you want anonymous communication with someone else, you should use anonymous channels such as Crows [6], MIX-net [7] and Onion routing [8] rather than *_anon key-exchange suites. Otherwise, some clues of your identity would leak out from the traffic. For example, IP headers include both source and destination addresses, which can be used to guess both the application and the user. *_anon can be used in circumstances where the server and client do not host certificates and require a confidential communication channel such as provided by WTLS class 1.

RSA and ECDH_ECDSA have the authentication mechanism based on the digital certificates. The RSA key exchange suite employs RSA public-key cryptosystem to transport keys, and RSA signature to issue the digital certificates. ECDH_ECDSA employs ECDH to exchange keys, and ECDSA to issue the certificates. Note that ECDH and ECDSA are the elliptic curve versions of the Diffie-Hellman key-exchange and the DSA signature (DSS), respectively. The transactions of RSA and ECDH_ECDSA are described as follows.

RSA Key Exchange Suite

1. A WAP gateway sends its certificate to a connecting client (a WAP device). Note that the certificate includes the gateway's RSA public-key and it is signed with RSA by a CA.

2. The client verifies whether or not the certificate is signed by a CA that the device trusts. If the signature is invalid or not signed by a trustful party, it aborts the communication (or alerts to the user).

3. If the certificate is valid, the client generates a secret value (used as the premaster secret), encrypts it with the gateway's public key and sends it to the gateway.

4. Both parties calculate their common keys using the shared secret value. If the client is to be authenticated it signs messages sent during the handshake with its RSA private key and sends the signature with its certificate.

ECDH_ECDSA Key Exchange Suite:

1. A WAP gateway sends its certificate to a connecting client (a WAP device). The certificate includes the gateway's ECDH public key and it is signed with ECDSA by a CA.

2. If the client is not to be authenticated, the client generates a temporal ECDH public key and sends it to the gateway. Otherwise, it sends its certificate, which includes its ECDH public-key and is signed with ECDSA by a CA.

3. Both parties calculate their common keys using the shared ECDH common key.

All the key-exchange suites starting with "EC" use elliptic curves, and each public-key includes a point on an elliptic curve with some parameters. A point on an elliptic curve is represented by a pair of x-coordinate and y-coordinate over a field, which is usually either a binary extension field or a prime field. The y-coordinate, however, can be compressed to one bit since once x-coordinate is given only two candidates remain for the y-coordinate. If "umcomp" is not in the name of the key-exchange suite the compressed form is used. Otherwise both x-coordinate and y-coordinate are used to express the point.

A number in a suite name, e.g. 768 of RSA_768, means the allowed maximum key size. If there is no number in it, it means no limitation on the key size. Limitations mainly come from the export restrictions. The Wassenaar Arrangement [40] requests the exporting entities to report to the government when exporting RSA or DH employing a key length in excess of 512 bits and ECDH of a key length in excess of 112 bits even though there are some exceptions. Please ask your government for more details. For security reasons, one should avoid RSA moduli smaller than 1024 bits, both DH and DSA keys smaller than 1024 bits, and then both ECDH and ECDSA keys smaller than 160 bits. As of the year 2003, RSA moduli up to 576 bits and ECDLP over a curve with the field size up to 109 bits were solved. For the current status, see [9, 10], respectively.

7.4.2   Elliptic Curve Parameters in WTLS

Some of the key exchange suites in WTLS use elliptic curves and WAP defines 12 elliptic curves. They are summarized in Table 7.3.

In Table 7.3, "yes" in the "Basic" column means that the corresponding curves must be implemented on all the WAP devices as long as they support elliptic curves. For the interoperability, they should be implemented on all the WAP gateways. "Characteristic" column shows the characteristic of the definition field of the elliptic curve. In it, "2" denotes a binary extension field and "p" denotes a prime field. If "(K)" follows "2," it means a Koblitz curve. Over a Koblitz curve, scalar multiplications of a point become faster using Frobenius mapping even though it also speeds up the attack by a

Table 7.3
Elliptic curves in WAP

Number	Basic	Field Size	Characteristic	Standardization
1		113	2 (K)	
2	unassigned			
3		162	2 (K)	ANSI X9.62-1, X9.63, FIPS186-2(K-163), SEC2(sect163k1)
4		133	2	
5	yes	163	2	
6		112	p	
7	yes	160	p	
8		112	p	
9		160	p	
10		223	2 (K)	ANSI X9.62-1, X9.63, FIPS186-2(K-223), SEC2(sect223k1)
11		223	2	ANSI X9.62-1, X9.63, FIPS186-2(B-223), SEC2(sect223r1)
12		224	p	

factor of $(2m)^{1/2}$ where m is the field size. Thus to balance the security level, the field size of a Koblitz curve should be larger than that of a random curve, i.e. the other curve than Koblitz, by around log2 m.

"Field Size" column denotes the size of the definition field of the elliptic curve. The field size is often referred to as a security parameter since it is usually only a little bit larger than the order size, which precisely depends on the complexity of the fastest general-purpose algorithm for solving ECDLP. Currently, the distributed version of Pollard's rho algorithm [11] is the fastest to solve ECDLP. As of the year 2003, ECDLP over a curve with the field size up to 109 was solved. See [10] for the current status. For a security reason, curves with field size smaller than 160 bits must not be used while there may be some regulations on the size, (e.g., for export). Actually, the Wassenaar Arrangement [40] requests the exporting entities to report to the government when exporting elliptic curve cryptosystems employing a key length in excess of 112 bits even though there are some exceptions. Please ask your government for more details. There is no restriction on signature algorithms since their purpose is to provide authentication and nonreputation but not to provide confidentiality (while it is not impossible to use them for confidentiality.) Elliptic curve #2 is unassigned, and then #3, #10, and #11 are standardized in ANSI X9.62-1 [12], X9.63 [13], FIPS186-2 [14], and SEC2 [15].

7.4.3 Bulk Encryption and MAC Suite

A shared key obtained by a key exchange suite is then used for bulk encryption/ decryption and MAC generation/verification. The available bulk ciphers and MAC parameters are listed in Tables 7.4 and 7.5, respectively.

In Table 7.4, NULL means no encryption. This option may be used if no encryption is required while both server authentication and data integrity are required. For example, if a server publishes information to the public no encryption is required while clients may request both authentication of the server and integrity of the downloading data.

DES [16], 3DES [17], and IDEA [18] are block ciphers of 64-bit block size. RC5 is also a block cipher but of a flexible block size

Table 7.4
Bulk Ciphers in WAP

Cipher	Assigned Number	Effective Key Size (Bits)
NULL	0	0
RC5_CBC_40	1	40
RC5_CBC_56	2	56
RC5_CBC	3	128
DES_CBC_40	4	40
DES_CBC	5	56
3DES_CBC_EDE	6	168
IDEA_CBC_40	7	40
IDEA_CBC_56	8	56
IDEA_CBC	9	128
RC5_CBC_64	10	64
IDEA_CBC_64	11	64

while 64-bit block is used in WTLS. DES is an obsolete standard and its key size is up to 56 bits. As of the year 2003, the key sizes up to 64 bits are broken with exhaustive search using globally available computational power [19]. Thus, DES and the other ciphers using smaller key sizes than 80 bits should not be used (80-64=16 bits are for security margin), even though there are some restrictions on export. The Wassenaar Arrangement [40] requests the exporting entities to report to the government when exporting symmetric algorithms of key length in excess of 56 bits. Please ask your government for more details. 3DES repeats DES three times in EDE (Encryption Decryption Encryption) mode using three distinct keys, k1, k2 and k3. Note that there are two kinds of 3DES, three-key 3DES and two-key 3DES. The two-key 3DES uses k1, k2 and then k1 again. The key length of the three-key 3DES is 168 bits, but its effective length is below 128 bits [20, 21]. In addition, 3DES is around three times slower than DES. Thus 3DES is not an efficient cipher. Both RC5 and IDEA support 128 bit key without decreasing encryption/decryption speed compared with DES. IDEA is patented in the United States [22]. CBC is an operation mode of a block cipher to encrypt a longer message than the block size. CBC mode can hide the dependency among the message blocks.

For the integrity check, WTLS employs HMAC [23], which uses a hash function SHA-1 [24] or MD5 [25], twice. Table 7.5 shows the available parameters for them. There are two general ways to forge a MAC. One is to exhaustive search the MAC key and then to forge the MAC. The other is to insert a random MAC without cracking the MAC key. The risk to the former attack is negligible if the MAC key is large enough. 16 byte key, i.e. 128 bit key, is secure enough in practice. The success probability of the latter attack is around $q/2^t$ where t is the MAC size and q is the total number of MACs an adversary can insert. Since q is not so large, $t = 80$ would be enough. The sizes in practice, however, should be chosen according to the required security level and performance. When the MAC algorithm is used as a pseudo random function(PRF) the MAC size is ignored and the full output size of the underlying hash function is used. The full size of SHA-1 is 20 bytes and that of MD5 is 16 bytes.

7.5 WAP-Profiled TLS

WAP2 follows the WAP-profiled TLS [26] that specifies cipher suites, session resume, session identifier, server authentication, client

Table 7.5
MAC Parameters in WAP

Hash Function	Assigned Number	Key Size (bytes)	MAC Size (bytes)
SHA_0	0	0	0
SHA_40	1	20	5
SHA_80	2	20	10
SHA	3	20	20
N/A	4	N/A	N/A
MD5_40	5	16	5
MD5_80	6	16	10
MD5	7	16	16

authentication and tunneling in accordance with the usual TLS 1.0 [27]. The features are summarized as follows.

7.5.1   Cipher Suites

A cipher suite is a suite of a key-establishment protocol, a bulk cipher and a MAC. Unlike WTLS (where any combinations of them are available), some combinations of them are predetermined. For example, "TLS_RSA_WITH_ DES_CBC_SHA" means RSA public-key cryptosystem for key establishment, DES in CBC mode for bulk encryption and hash function SHA-1 for MAC calculation. Table 7.6 lists up all the cipher suites defined in TLS 1.0, but additional cipher suites will be defined with another RFCs. They will most likely include AES block cipher, public-key cryptosystems and signatures over elliptic curves.

"TLS_NULL_WITH_NULL_NULL" does not provide any security, and thus must not be used not only for mission critical but any communication that requires a minimum level of security. Both "TLS_RSA_WITH_NULL_MD5" and "TLS_RSA_WITH_NULL_ SHA" provide no encryption while the server is identified using RSA and then authentic channels are established to the server using MD5 and SHA-1, respectively. More precisely, the server has its RSA public-key, which is signed by a CA using RSA signature with the server's ID. The client verifies the CA's signature, encrypts a secret with the verified public-key and sends it to the server. The server decrypts it and shares the secret with the client. Using the shared secret, both the client and the server establish authentic channels, in which modification is detectable using MAC algorithm. This suite is useful when a server publishes information to the public. Since the information is public, no encryption is required while clients may require both authentication of the publishing server and the integrity of the downloading data.

The other suites support encryption of the communication channel in addition to the key-establishment and the integrity check over the established channels. Thus they are useful for making encryption tunnels between a client and an authenticated sever. The tunnel can then be used, for instance, for forwarding a user's password so that the server can authenticate the user and then can display his/her personalized pages that may include sensitive information.

Table 7.6
Cipher Suites in TLS 1.0

Cipher Suite	Assigned Number
TLS_NULL_WITH_NULL_NULL	{ 0x00,0x00 }
TLS_RSA_WITH_NULL_MD5	{ 0x00,0x01 }
TLS_RSA_WITH_NULL_SHA	{ 0x00,0x02 }
TLS_RSA_EXPORT_WITH_RC4_40_MD5	{ 0x00,0x03 }
Ii0TLS_RSA_WITH_RC4_128_MD5	{ 0x00,0x04 }
TLS_RSA_WITH_RC4_128_SHA	{ 0x00,0x05 }
TLS_RSA_EXPORT_WITH_RC2_CBC_40_MD5	{ 0x00,0x06 }
TLS_RSA_WITH_IDEA_CBC_SHA	{ 0x00,0x07 }
TLS_RSA_EXPORT_WITH_DES40_CBC_SHA	{ 0x00,0x08 }
TLS_RSA_WITH_DES_CBC_SHA	{ 0x00,0x09 }
TLS_RSA_WITH_3DES_EDE_CBC_SHA	{ 0x00,0x0A }
TLS_DH_DSS_EXPORT_WITH_DES40_CBC_SHA	{ 0x00,0x0B }
TLS_DH_DSS_WITH_DES_CBC_SHA	{ 0x00,0x0C }
TLS_DH_DSS_WITH_3DES_EDE_CBC_SHA	{ 0x00,0x0D }
TLS_DH_RSA_EXPORT_WITH_DES40_CBC_SHA	{ 0x00,0x0E }
TLS_DH_RSA_WITH_DES_CBC_SHA	{ 0x00,0x0F }
TLS_DH_RSA_WITH_3DES_EDE_CBC_SHA	{ 0x00,0x10 }
TLS_DHE_DSS_EXPORT_WITH_DES40_CBC_SHA	{ 0x00,0x11 }
TLS_DHE_DSS_WITH_DES_CBC_SHA	{ 0x00,0x12 }
TLS_DHE_DSS_WITH_3DES_EDE_CBC_SHA	{ 0x00,0x13 }
TLS_DHE_RSA_EXPORT_WITH_DES40_CBC_SHA	{ 0x00,0x14 }
TLS_DHE_RSA_WITH_DES_CBC_SHA	{ 0x00,0x15 }
TLS_DHE_RSA_WITH_3DES_EDE_CBC_SHA	{ 0x00,0x16 }
TLS_DH_anon_EXPORT_WITH_RC4_40_MD5	{ 0x00,0x17 }
TLS_DH_anon_WITH_RC4_128_MD5	{ 0x00,0x18 }
TLS_DH_anon_EXPORT_WITH_DES40_CBC_SHA	{ 0x00,0x19 }
TLS_DH_anon_WITH_DES_CBC_SHA	{ 0x00,0x1A }
TLS_DH_anon_WITH_3DES_EDE_CBC_SHA	{ 0x00,0x1B }

The tunnel may also be used to submit a credit card number for web shopping. (submission of credit card numbers, however, need to be handled with care, since the submitted data are decrypted at the server and might be abused if the server is not trustworthy.)

The suites including DH_anon and/or EXPORT should be avoided if middle or high level of security is required. As mentioned in the key exchange suite of WTLS, DH_anon is anonymous (non-authentic) Diffie-Hellman key-exchange, which is vulnerable to the intruder-in-the-middle attack. EXPORT means exportable from the U.S. and uses 40-bit symmetric encryption key and up to 512-bit RSA or DH keys. The regulation was, however, relaxed to almost nothing while the Wassenaar Arrangement [40] still requests that the exporting entities should report to the government when exporting symmetric encryption algorithms of key length in excess of 56 bits, RSA and DH of key length in excess of 512 bits, and elliptic curve cryptosystems of key length in excess of 112 bits even though there are some exceptions. Please ask your government for more details. A 40-bit symmetric key is exhaustible within a couple of days using several PCs [19] and a 512-bit RSA key can be cracked around half a year using hundreds of workstations and PCs [9].

Explanation of the other components is given as follows. DH denotes the ephemeral-static Diffie-Hellman key exchange where the client's DH public-key is temporal and varies every time and the server's DH public key is fixed and certified by a CA usually. DH_RSA and DH_DSS further specify the signing algorithm for the certificate of the server's public key. DHE denotes ephemeral Diffie-Hellman where both the client's and the server's DH public keys are temporal and the server signs its DH public key using RSA or DSS according to DHE_ RSA or DHE_ DSS. The signing algorithm for the server's certificate is specified by the DHE parameter. 3DES_EDE denotes three-key triple DES and EDE means encryption-decryption-encryption process in the triple DES.

WAP TLS profile specifies that WAP servers must support both "TLS_RSA_WITH_RC4_128_SHA" and "TLS_RSA_WITH_3DES_EDE_CBC_SHA," and that WAP clients must support at least one of them. Of course, they may support any other cipher suites including new ones. The difference between the two specified suites is whether RC4 or 3DES for the bulk encryption. If the encryption speed is important RC4 would be better, otherwise 3DES. RC4 is a byte-oriented stream cipher designed by R.L. Rivest. Its algorithm was originally kept in secret by RSA Data Security (currently RSA Security), but someone reverse engineered it and posted its source code on the Internet. Currently, it is available anywhere, [28], and is one of the most famous and widely used ciphers. While RC4 is faster than DES (and of course than 3DES), it has some undesirable properties. One is that the output stream of RC4 has biases in its initial bytes [29, 30]. (It is recommended to discard the first 256 to 512 bytes of it or to encrypt garbage first.) The other is that the knowledge of a part of a key may be used to guess the other part of the key [31, 32]. This is critical in WEP [33] since in WEP the initial three bytes of the RC4 key are open to the public as the IV (Initial Value). The three byte IV varies every packet and the latter key is fixed. By collecting enough packets including weak IVs, an adversary can crack the fixed key. Fortunately, this attack cannot be applied to the RC4 in TLS (and in TKIP [34]) since any part of a key is not open to the public and the key is replaced every a certain period.

7.5.2 Session ID and Session Resume

Session identifier (Session ID) is a byte sequence chosen by the server. It is used to identify a particular session, which consists of a client certificate (if the client is to be authenticated), compression method, cipher spec, 48-byte master_secret and a flag whether or not the session is resumable. While TLS 1.0 accepts an arbitrary byte sequence as an ID (SSL version 2 uses 16 byte ID), WAP TLS profile recommends 8 byte or less IDs to improve over the air efficiency.

If the flag of the session resumable is true and both the client and the server agree on the resume, the closed previous session can be resumed. It is very useful for low-computational-power terminals since the session resume skips heavy key-establishment operations. A full key-establishment uses asymmetric-key cryptographic operations, such as RSA, Diffie-Hellman key-exchange and DSA, which are roughly a thousand times as heavy as symmetric-key cryptographic operations, such as DES, SHA-1, MD5 and HMAC. Thus the session resume is useful for low-computational-power terminals.

While a resumed session keeps on using the previous master_secret, the encryption keys and the MAC keys are updated with a simplified key-establishment using the master_secret. In the simplified key-establishment, both parties exchange ClientHello.random and ServerHello.random, hash them with the session's master_secret and then generate new encryption and MAC keys. The advantage of updating keys is that the updated keys cannot be guessed even if the old ones are compromised thanks to the one-wayness of the hash function. The compromise of the master_secret, however, breaks all the consequent encryption and MAC keys. The master_secret is updated with the full key-establishment.

TLS 1.0 suggests limiting a session life up to 24 hours and WAP TLS profile recommends a longer session life, about 12 hours.

7.5.3   Server/Client Authentication and Certificate

In the same way as TLS 1.0, the server authentication is mandatory and the client authentication is optional in the WAP profile of TLS. The authentication in TLS is a certification base, in which one verifies a given certificate and then checks whether the communicating entity has the private-key corresponding to the public-key written in the certificate. A valid certificate tells us the issuer (usually CA) of it and the owner ID of the public-key in the certificate. The correctness of the owner ID, however, depends on the trust level of the issuer and then the level must be evaluated by each user themselves. (This is the hardest part of PKI and the other options are introduced in [35, 36].)

Anyway, servers, clients and issuers supporting WAP must follow "WAP Certificate and CRL Profile" [4], and should follow the usual X.509 certificate specification [5].

7.5.4   TLS Tunneling

As mentioned in Section 7.2, a WAP proxy may be placed in between a client and a server to adapt the TCP profile to the wireless communication. The proxy simply functions as a relay node in the transport level and is isolated from the TLS session unlike the proxy in WAP1.0 that breaks a secure tunneling there. The WAP profile of TLS maintains the end to end security at the transport layer even though it requires the client to support the TLS tunneling by itself.

7.6 WAP Identity Module

WIM (WAP Identity Module) is an identity module for WAP. It provides identity of a user over the network. WAP separates its identification functionality from WAP devices similarly to the relationship between the GSM mobile phones and the subscriber identity modules. This separation enables users to update their devices without changing their telephone numbers and their billing properties. This also enables bearers not to give keys for identification to device manufactures. WIM is usually implemented on a smart card where the memory size and the computational power are limited. WIM applications are designed so that they can be implemented within the applications. WIM takes on the operations such as the following:

- (Generation of a key pair) and storing the private-key;
- Generation of cryptographically secure random numbers;
- Signing operation using both the stored private-key and the generated random numbers;
- Decryption operation using the stored private-key;
- Key-exchange operation using the stored private-key or an ephemeral private-key;
- Storing self-certificates of trustful CAs;
- Storing a certificate chain or the CertificateURL of the module.

The upper five operations require confidentiality, that is, the inner data must not be monitored by outside adversaries. In addition, the signing operation requires non-repudiation, which is the property that only the user can perform the operation. This property is enhanced by protecting the signing key even against the owner so that the owner cannot reveal it to the public and claim that it is cracked. On the contrary, public-key encryption and signature verification can be performed outside WIM since they do not require private-key operations.

The sixth operation requires data integrity, i.e. modification or replacement of the installed certificates should be detectable. If an adversary can install his self-certificate into a target user's WIM as a trustful one, the adversary can cheat the user into accepting his fake server as a genuine server and then can get the user's password and/or credit card numbers. Note that the confidentiality is not required since certificates are public.

Storing a certificate chain or a CertificateURL of the WIM is just for the portability of the user. Both exposure and modification of them are not a matter since they are open to the public and the modification can be detected using the issuer's certificate. If the size of them is large, they may be placed on a memory outside of the WIM. Actually, CertificateURL is for referring to the outside DB.

Temporal keys, such as MAC keys and bulk decryption keys, may also be stored outside WIM, but in a WAP device. Since WAP devices have more memory and more computational power, this increases the processing speed. Leakage of temporal keys is not a big issue since they have a lifetime negotiated during the handshake and then deleted usually within 24 hours.

7.7 Further Information

As of 2004, WAP specifications are maintained by the Open Mobile Alliance (OMA) Ltd. [37] while WAP was originally maintained by the WAP Forum that was an open forum founded in 1997. OMA was created in June, 2002 by consolidating the WAP Forum and the Open Mobile Architecture Initiative (OMAI), another group for open mobile architecture, which was established in November 2001.

The latest security related issues and specifications are discussed in OMA, e.g. in the security working group [38] and in the mobile-commerce and charging working group [39]. The security working group covers the topics like:

- Protocols for secure communication between mobile clients and servers;

- Addressing security related issues regarding new services being defined within in OMA (device management, messaging, Push services and the like);

- Interactions with other devices, such as hardware tokens and smart cards.

And the mobile-commerce and charging working group covers:

- A consistent interface to facilitate charging;
- Payment and charging systems that would be interoperable with other specifications and that can be achieved over the OMA architecture.

Another hot topic is digital right management(DRM) of download contents, such as melodies as calling signals, background images, Java applications, music and streaming movies. Current status is available from the web pages of OMA.

References

[1] Fielding, R., et. al., "Hypertext Transfer Protocol—HTTP/1.1," January 1997, ftp:// ftp.isi.edu/in-notes/rfc2068.txt.

[2] Khare, R., and S. Lawrence, "Upgrading to TLS Within HTTP/1.1," May 2000, http://www.ietf.org/rfc/rfc2817.txt.

[3] Rescorla, E., "HTTP over TLS," May 2000, http://www.ietf.org/rfc/rfc2818.txt.

[4] "WAP Certificate and CRL Profiles," *WAP-211-WAPCert, Draft Version,* May 22, 2001, WAP Forum, http://www.openmobilealliance.org/tech/affiliates/wap/ wapindex.html.

[5] Housley, R., et al., "Internet X.509 Public Key Infrastructure Certificate and CRL Profile," rfc 2459, January 1999, http://www.ietf.org/rfc/rfc2459.txt.

[6] Reiter, M., and A. Rubin, "Crowds: Anonymity for Web Transactions," *DIMACS Technical Report,* Vol. 97, No. 15, April 1997.

[7] Chaum, D., "Untraceable Electronic Mail, Return Addresses, and Digital Pseud-onyms," *Communications of the ACM,* Vol. 4, No. 2, February 1981.

[8] Syverson, P., M. Reed, and D. Goldschlag, "Onion Routing Access Configura-tions," *DARPA Information Survivability Conference and Exposition (DISCEX 2000),* Vol. 1, 2000, pp. 34–40.

[9] "The New RSA Factoring Challenge," http://www.rsasecurity.com/rsalabs/ challenges/factoring/index.html.

[10] "The Certicom ECC Challenge," http://www.certicom.com/index.php?action= res,ecc_ challenge.

[11] Oorschot, P. V., and M. Wiener, "Parallel Collision Search with Cryptanalytic Applications," *Journal of Cryptology*, Vol. 12, 1999, pp. 1–28.

[12] "Public Key Cryptography for the Financial Services Industry: The Elliptic Curve Digital Signature Algorithm (ECDSA)," ANSI X9.62-1999, 1999.

[13] "Public Key Cryptography for the Financial Services Industry: Key Agreement and Key Transport Using Elliptic Curve Cryptography," ANSI X9.63-200x Working Draft, October 2000.

[14] "Digital Signature Standard," NIST FIPS PUB 186-2 (+Change Notice), National Institute of Standards and Technology, U.S. Department of Commerce, January 2000, http://csrc.nist.gov/publications/fips/fips186-2/fips186-2-change1.pdf.

[15] "SEC 2: Recommended Elliptic Curve Domain Parameters," *Standards for Efficient Cryptography Group*, Version 1.0, September 2000, http://www.secg.org/.

[16] "American National Standard for Information Systems—Data Link Encryption," *ANSI X3.106*, American National Standards Institute, 1983.

[17] Tuchman, W., "Hellman Presents No Shortcut Solutions To DES," *IEEE Spectrum*, Vol. 16, No. 7, July 1979, pp 40–41.

[18] Lai, X., "On the Design and Security of Block Ciphers," *ETH Series in Information Processing*, Vol. 1, Konstanz: Hartung-Gorre Verlag, 1992.

[19] "The RSA Data Security Secret-Key Contests," http://www.rsasecurity.com/ rsalabs/challenges/secretkey/secret-key.html.

[20] Merkle, R. C., and M. Hellman, "On the Security of Multiple Encryption," *Communications of the ACM*, Vol. 24, No. 7, 1981, pp. 465–467.

[21] Lucks, S., "Attacking Triple DES," *Proc. of Fast Software Encryption '98*, LNCS 1372, 1998, pp. 239–253.

[22] "Device for the conversion of a Digital Block and Use of Same," U.S. Patent, 5214703.

[23] Krawczyk, H., M. Bellare, and R. Canetti, "HMAC: Keyed-Hashing for Message Authentication," February 1997, ftp://ftp.isi.edu/in-notes/rfc2104.txt.

[24] "Secure Hash Standard," NIST FIPS PUB 180-1, National Institute of Standards and Technology, U.S. Department of Commerce, DRAFT, May 1994.

[25] Rivest, R., "The MD5 Message Digest Algorithm," RFC 1321, April 1992. URL: ftp:// ftp.isi.edu/in-notes/rfc1321.txt.

[26] "WAP TLS Profile and Tunneling," WAP-219-TLS, Draft Version 11, April-2001, WAP Forum, http://www.openmobilealliance.org/tech/affiliates/ wap/wapindex.html.

[27] Dierks, T., and C. Allen, "The TLS Protocol, Version 1.0" rfc 2246, January 1999, http://www.ietf.org/rfc/rfc2246.txt.

[28] Schneier, B., *Applied Cryptography: Protocols, Algorithms and Source Code in C.*, New York: John Wiley and Sons, 2nd ed., 1996.

[29] Fluhrer, S., and D. McGrew, "Statistical Analysis of the Alleged RC4 Keystream Generator," *Proc. of Fast Software Encryption 2000*, LNCS 1978, 2000, pp. 19–30.

[30] Mironov, I., "Statistical Analysis of the Alleged RC4 Keystream Generator," *Proc. of CRYPTO 2002*, LNCS 2442, 2002, pp. 304–319.

[31] Fluhrer, S., I. Mantin, and A. Shamir, "Weaknesses in the Key Scheduling Algorithm of RC4," *SAC'01*, LNCS 2259, 2001, pp. 1–24.

[32] Stubblefield, A., J. Ioannidis, and A. Rubin, *Using the Fluhrer, Mantin, and Shamir Attack to Break WEP*, ATT Labs Technical Report, TD4ZCPZZ, Revision 2, 2001.

[33] Wireless LAN Medium Access Control (MAC) and Physical Layer (PHY) Specifications, IEEE Std 802.11, 1999 Edition, http://standards.ieee.org/reading/ieee/std/lanman/.

[34] Wireless LAN Medium Access Control (MAC) and Physical Layer (PHY) Specifications: Specification for Enhanced Security, IEEE P802.11i/D3, http://standards.ieee.org/reading/ieee/std/lanman/.

[35] Kobara, K., and H. Imai, "Pretty-Simple Password-Authenticated Key-Exchange Protocol Proven to Be Secure in the Standard Model," *IEICE Trans.*, E85-A (10), October 2002, pp. 2229–2237.

[36] Shin, S., H., K. Kobara, and H. Imai, "Leakage-Resilient Authenticated Key Establishment Protocols," *Proc. of ASIACRYPT 2003*, LNCS 2894, Springer-Verlag, 2003, pp. 155-172.

[37] Open Mobile Alliance, http://www.openmobilealliance.org/.

[38] OMA Security Working Group, http://www.openmobilealliance.org/tech/wg_committees/sec.html.

[39] OMA M-Commerce and Charging Working Group, http://www.openmobilealliance.org/tech/wg_committees/mcc.html.

[40] "The Wassenaar Arrangement," http://www.wassenaar.org/.

About the Authors

Hideki Imai is a professor at the University of Tokyo and currently serves as the director of the Research Center for Information Security at the National Institute of Advanced Industrial Science and Technology. He received a B.E., M.E., and Ph.D. in electrical engineering from the University of Tokyo in 1966, 1968, and 1971, respectively. From 1971 to 1992 he was on the faculty of Yokohama National University. In 1992 he joined the faculty of the University of Tokyo, where he is currently a professor in the Institute of Industrial Science. His current research interests include information theory, coding theory, cryptography, and information security. Dr. Imai has received many awards, in addition to official commendations from the Minster of Internal Affairs and Communications and from the Minister of Economy, Trade, and Industry. Dr. Imai was awarded an honorary doctor degree by Soonchunhyang University, Korea, in 1999 and Docteur Honoris Causa by the University of Toulon Var, France, in 2002. He was elected an IEEE Fellow in 1992 and an IEICE Fellow in 2001. He chaired many committees of scientific societies and organized a number of international conferences. He served as the president of the Society of Information Theory and its Applications in 1997, the IEICE Engineering Sciences Society in 1998, and the IEEE Information Theory Society in 2004. Dr. Imai is currently the chair of CRYPTREC (Cryptography Techniques Research and Evaluation Committee of Japan).

Mohammad Ghulam Rahman is a postdoctoral fellow at the University of Calgary in Canada. He obtained a B.Sc. in electrical and electronic engineering from Bangladesh University of Engineering and Technology (BUET) in 1993. He received an M.S. in telecommunication from the Asian Institute of Technology (AIT), Thailand in 1998. In 2003 he received a Ph.D. in information and communication engineering from the University of Tokyo, Japan. He worked as a research fellow in National Institute of Information and Communication Technology (NiCT), Japan. Currently, he is a post- doctoral fellow at the University of Calgary, Canada. His research interests are wireless communication, cryptography and its application, data security, and mobile commerce.

Kazukuni Kobara is a research associate at the University of Tokyo and currently serves as the team leader of the Research Team for Security Fundamentals in the Research Center for Information Security at the National Institute of Advanced Industrial Science and Technology. In 1994 he joined the Institute of Industrial Science of the University of Tokyo and received a Ph.D. in engineering from the University of Tokyo in 2003. His current research interests include cryptography, information security, and network security. Dr. Kobara received the SCIS Paper Award and the Vigentennial Award from the ISEC group of IEICE in 1996 and 2003, respectively. He also received the Best Paper Award of WISA in 2001, the ISITA Paper Award for Young Researchers in 2002, and the IEICE Best Paper Award (Inose Award) in 2003. He is a member of the IEICE of Japan and IACR. He served as a member of CRYPTREC (2000 to the present) and as the vice-chairperson of the WLAN security committee in 2003.

Anderson C. A. Nascimento works for NTT Laboratories in Yokosuka, Japan. He received a Ph.D. from the University of Tokyo in 2004.

Tatsuro Oi is an engineer at NTT DoCoMo, Inc. He joined NTT DoCoMo in 1999 and was involved in the development of i-mode mobile phones. He is currently with the Customer Equipment Development Department at NTT DoCoMo.

Index

The Artech House Universal Personal Communications Series

Ramjee Prasad, Series Editor

WLAN Systems and Wireless IP for Next Generation Communications,
Neeli Prasad and Anand Prasad, editors

WLANs and WPANs towards 4G Wireless, Ramjee Prasad and
Luis Muñoz

For further information on these and other Artech House titles,
including previously considered out-of-print books now available
through our In-Print-Forever® (IPF®) program, contact:

Artech House	Artech House
685 Canton Street	46 Gillingham Street
Norwood, MA 02062	London SW1V 1AH UK
Phone: 781-769-9750	Phone: +44 (0)20 7596-8750
Fax: 781-769-6334	Fax: +44 (0)20 7630-0166
e-mail: artech@artechhouse.com	e-mail: artech-uk@artechhouse.com

Find us on the World Wide Web at: www.artechhouse.com